MICELLES, MONOLAYERS, AND BIOMEMBRANES

MICELLES, MONOLAYERS, AND BIOMEMBRANES

MALCOLM N. JONES
School of Biological Sciences
University of Manchester
Manchester, United Kingdom

DENNIS CHAPMAN
Royal Free Hospital School of Medicine
London, United Kingdom

WILEY-LISS

A JOHN WILEY & SONS, INC., PUBLICATION
New York • Chichester • Brisbane • Toronto • Singapore

Address All Inquiries to the Publisher
Wiley-Liss, Inc., 605 Third Avenue, New York, NY 10158-0012

© 1995 Wiley-Liss, Inc.

Printed in the United States of America

While the authors, editors, and publisher believe that drug selection and dosage and the specifications and usage of equipment and devices, as set forth in this book, are in accord with current recommendations and practice at the time of publication, they accept no legal responsibility for any errors or omissions, and make no warranty, express or implied, with respect to material contained herein. In view of ongoing research, equipment modifications, changes in governmental regulations and the constant flow of information relating to drug therapy, drug reactions and the use of equipment and devices, the reader is urged to review and evaluate the information provided in the package insert or instructions for each drug, piece of equipment or device for, among other things, any changes in the instructions or indications of dosage or usage and for added warnings and precautions.

Library of Congress Cataloging-in-Publication Data

Jones, Malcolm N.
    Micelles, monolayers, and biomembranes / by Malcolm N. Jones and
Dennis Chapman.
      p.   cm.
    Includes bibliographical references and index.
    ISBN 0-471-56139-8.   —   ISBN 0-471-30596-0 (pbk.)
    1. Membranes (Biology) 2. Hydrophobic surfaces. 3. Micelles.
4. Monomolecular films. I. Chapman, Dennis, 1927–   . II. Title.
QH601.J614   1994
574.87'5—dc20
                                              94-30352
                                                CIP

The text of this book is printed on acid-free paper.

CONTENTS

PREFACE   xi

1. THE CHEMISTRY AND ROLE OF AMPHIPATHIC
   MOLECULES   1

   1.1. Occurrence of Amphipathicity at the Molecular Level / 1
   1.2. Types of Amphipathic Molecules / 5
         1.2.1. Synthetic Surfactants / 5
         1.2.2. Natural Surfactants / 8
         1.2.3. Membrane Lipids / 10
         1.2.4. Amphipathic Nature of Proteins / 13
   1.3. The Role of Amphipathic Molecules in
         Biomembranes / 17
   1.4. The Role of Amphipathic Structures in Protein
         Targeting / 20

2. MONOLAYERS   24

   2.1. The Formation of Monolayers / 24
   2.2. Techniques Used to Study Monolayers at Liquid
         Interfaces / 25
   2.3. States of Monolayers / 29
   2.4. Phospholipid Monolayers / 31
   2.5. Phospholipid Monolayer to Surface Bilayer
         Transition / 35

2.6.   Ion Interactions and Electrostatic Effects in Phospholipid
       Monolayers / 36
2.7.   Mixed Monolayers / 39
2.8.   Adsorption of Polypeptides and Proteins at the Aqueous–
       Air Interface / 45
       2.8.1.   Polypeptide Monolayers / 45
       2.8.2.   Protein Monolayers / 50
                2.8.2.1.   Kinetics of adsorption / 50
                2.8.2.2.   Structure of surface layers / 54
                2.8.2.3.   Protein lipid interactions in
                           monolayers / 56
2.9.   Equivalence of Monolayers and Bilayers / 58

3. THE MICELLAR STATE   64

3.1.   The Critical Micelle Concentration and Micelle
       Formation / 64
       3.1.1.   The cmc of Synthetic Surfactants and the Effects
                of Composition of the Medium / 67
       3.1.2.   The cmc of Natural Surfactants / 71
       3.1.3.   The cmc of Phospholipids / 74
       3.1.4.   Micellar-like Behavior of Some Protein
                Associations / 76
3.2.   The Size and Shape of Micelles / 79
       3.2.1.   General Considerations / 79
       3.2.2.   Bile Salt Micelles / 81
       3.2.3.   Phospholipid Micelles / 84
       3.2.4. β-Casein Micelles / 86
3.3.   Thermodynamics of Micellization / 89
3.4.   Mixed Amphiphile Systems / 95
3.5.   Surfactants in the Environment / 98

4. LIPID HYDRATED STATES AND PHASE BEHAVIOR   102

4.1.   Lyotropic Mesomorphism / 102
4.2.   The Lamellar Phases / 103
       4.2.1.   The Order Profile Parameter / 107
       4.2.2.   Correlation With Monolayer Properties / 108
       4.2.3.   Effects of Cholesterol / 108
4.3.   Phase Separation / 110
4.4.   Hexagonal Phases / 112
4.5.   Cubic Phases / 113

5. THE LIPOSOMAL STATE   117

5.1. Liposomes / 117
5.2. The Formation of Liposomes / 117
 5.2.1. Types of Liposomes / 117
 5.2.2. Methods of Preparation / 118
  5.2.2.1. Multilamellar vesicles / 118
  5.2.2.2. Small unilamellar vesicles / 118
 5.2.3. The Stability of Liposomes / 120
5.3. Liposomes as Carriers / 121
5.4. Liposomal Targeting / 124
 5.4.1. Natural Targeting (Passive Targeting) / 124
 5.4.2. Physical Targeting / 124
 5.4.3 Compartmental Targeting / 126
 5.4.4. Ligand-Mediated Targeting (Active Targeting) / 126
  5.4.4.1. Preparation and characterization of proteoliposomes / 127
5.5. The Interaction of Liposomes With Cells / 134
5.6. The Problem of the RES in Drug Delivery by Liposomes / 136

6. SURFACANT INTERACTIONS WITH BILAYERS, MEMBRANES, AND PROTEINS   143

6.1. Surfactant-Membrane Interactions / 143
 6.1.1. Surfactant-Lipid Interactions / 144
 6.1.2. The Mechanism of Surfactant-Induced Membrane Solubilization / 147
6.2. Surfactant-Protein Interactions / 149
 6.2.1. Denaturing and Nondenaturing Surfactants / 150
 6.2.2. The Characterization and Mechanism of Surfactant Binding and Denaturation of Proteins / 151
6.3. Techniques in the Study of Surfactant-Protein Interaction / 153
6.4. Thermodynamics of Surfactant-Protein Interactions / 156
6.5. The Structure of Protein-Surfactant Complexes / 167
6.6. Molecular Dynamics of Protein-Surfactant Complexes / 172

7. MEMBRANE STRUCTURE   180

7.1. The Composition of Biomembranes / 180

    7.1.1. Lipids / 180

    7.1.2. Proteins / 183

7.2. The Origin of the Bilayer Concept in Membranes / 185

7.3. The Fluid-Mosaic Model / 185

7.4. Static and Dynamic Lipid Asymmetry in Cell Membranes / 186

7.5. Protein and Glycoprotein Arrangements / 187

    7.5.1. Integral Proteins / 188

    7.5.2. Extrinsic Proteins / 188

7.6. Ion Channel Structures and Receptor Proteins / 189

    7.6.1. Receptor Proteins / 189

7.7. Lipid Phase Transitions and Biomembrane Systems / 190

    7.7.1. The Hexagonal Phase / 191

7.8. The Structure of the Human Erythrocyte Membrane / 192

7.9. Spectroscopic Studies of Biomembranes / 195

    7.9.1. Bacteriorhodopsin and Rhodopsin / 196

8. RECONSTITUTION OF MEMBRANE FUNCTION IN BILAYER SYSTEMS   199

8.1. The Objectives of Reconstitution / 199

8.2. General Principles of Reconstitution / 200

8.3. Choice of Surfactant for Reconstitution Studies / 203

8.4. Reconstituted Membrane Protein/Glycoprotein Systems / 204

    8.4.1. The Role of Lipid in Membrane Transporter Function / 208

    8.4.2. The Role of Lipid in Membrane Receptor Function / 212

    8.4.3. The Effects of Cholesterol in Reconstitution / 214

9. PROTEIN TRANSLOCATION AND THE ANCHORING AND DISPOSITION OF PROTEINS   220

9.1. Protein Translocation Across Membranes / 220

9.2. Signal Peptides / 221

9.3. Chaperons / 222

9.4. Protein Insertion / 224

9.5. Lipid-Protein Interactions / 225

9.6. Anchoring of Peripheral (Extrinsic) Membrane
Proteins / 227

9.7. Covalent Attachment of Fatty Acids to Proteins / 228

10. MEMBRANE DYNAMICS   **232**

10.1. Molecular Motion of Biomembranes / 232

10.1.1. Lipid Lateral Diffusion / 232

10.1.2. Lipid Flip-Flop / 233

10.2. Molecular Motion of Membrane Proteins / 234

10.2.1. Lateral Diffusion of Proteins / 234

10.2.2. Rotational Diffusion of Membrane
Proteins / 237

10.2.3. Flippase Proteins / 238

10.3. Transmembrane Signaling and Receptor
Processing / 239

10.3.1. Transmembrane Signaling / 239

10.3.2. Endocytosis and Receptor Processing / 239

INDEX   **245**

PREFACE

The objective in writing this book was to attempt to present in an up-to-date single volume those phenomena that arise as a consequence of hydrophobicity. The formation of monolayers, micelles, liposomes, and natural membranes (biomembranes) is to varying degrees a direct consequence of the amphipathic nature of their constituent parts, and the properties of such systems are intimately affected by hydrophobic interactions. In bringing together a body of information on the nature of such systems we hope that the reader will be able to appreciate and to understand the immense importance of amphipathic molecules and their role in determining the structure and properties of monolayer, bilayer, and biomembrane systems.

We are of course conscious of the enormous size of the monolayer, micelle, and biomembrane fields; whole volumes could be devoted to each one of these areas. We have therefore had to be selective in the topics we have chosen to discuss. Such choices are difficult to make, and subjective. Inevitably we may have omitted areas that some will think might also have been included. We have nevertheless endeavoured to give a broad and up-to-date coverage of the fields where amphipathic molecules are involved. Thus we have also included chapters where such topics as protein translocation, chaperones, and biomembrane dynamics are discussed, as these are also directly related to the amphipathic character of these systems. We have included further reading suggestions in some of the chapters as well as specific references.

While the book assumes some knowledge of both physical chemistry and biochemistry/biology, we have attempted as far as possible to give appropriate introductory information, although of necessity this is brief in parts. We hope the book will be of general interest to third-year students of both the physical and biological sciences and to those researchers who wish to be introduced to the manifestations of hydrophobic interactions. Finally, we would like to express our appreciation to our typists, Miss Lynne Smith in Manchester and Mrs. Laks Sinha in London, for word processing technically difficult manuscripts, and to Mrs. Christine Hall in London for general assistance. Special thanks are also due to Mr. Steven Wilkinson and his father, Dr. Alan E. Wilkinson, who have redrawn in a unified fashion, using computer graphics, all the figures taken from the literature. Their work has greatly enhanced the appearance of the book. We also thank Mr. Jason W. Birkett in the Department of Chemistry at Manchester for drawing the organic formulae in Chapter 5.

MALCOLM N. JONES
DENNIS CHAPMAN

Manchester and London, 1994

CHAPTER 1

THE CHEMISTRY AND ROLE OF AMPHIPATHIC MOLECULES

1.1. OCCURRENCE OF AMPHIPATHICITY AT THE MOLECULAR LEVEL

The occurrence of molecules having both hydrophobic and hydrophilic moieties in chemical combination gives rise to an extensive range of phenomena and structural patterns, the importance of which can hardly be underestimated in both the physical and life sciences. The widespread use of soaps and detergents, the stabilization of foams and emulsions in the processing of beverages and foodstuffs, the stabilization of particulate dispersions such as paints, and the secondary recovery of oil from porous rock beds are all processes that depend on the use of molecules with a dual character, i.e., amphipathic molecules. It is almost inconceivable to consider the existence of life forms and living processes without amphipathic molecules. From the stabilization of globular proteins to the formation of complex organelles and membranous structures, amphipathicity plays a vital role in facilitating the development of the structural features necessary for the evolution and maintenance of life processes.

The simultaneous existence of chemical entities having an affinity and an antipathy for water within the same molecule is the pivot around which amphipathicity revolves. In terms of chemical structure, amphipathic molecules have hydrophobic alkyl, acyl, or aromatic groups in combination with polar and/or ionizing groups with hydrophilic character. It is the need to satisfy these conflicting characteristics and minimize the energy of the system that gives rise to the formation of a host of complex structural features involving the coming together of the hydrophobic residues in the bulk phase to form aggregated

1

structures such as micelles, and that leads to interfacial phenomena such as the formation of monolayers, bilayers, and related vesicular structures and, of course, the all-important existence of the biological membrane.

The manifestation of amphipathicity requires the existence of what is the most unique liquid on the planet: water. Water has been called the "matrix of life" (A. Szent-Gyorgy, 1957), and it is water's unique physical properties, together with its abundance, that has enabled life to be evolved in its present form. Table 1.1 shows some of the physical properties of water compared with those of three other common liquids, methanol, benzene, and n-hexane. Perhaps the most unusual feature of water is that it is a liquid at all at ambient temperatures. The "hydrides" of nitrogen and sulphur—ammonia (NH_3, molecular mass 17.03) and hydrogen sulphide (H_2S, molecular mass 34.08)—have melting points of $-77.7°C$ and $-85.5°C$, respectively, and are gaseous at room temperatures. Thus water has a high melting point and a high boiling point for its molecular mass; it also has an exceptionally high heat capacity, high surface tension, and high relative permittivity. These characteristics are a consequence of the structure of water in the solid and liquid states, which in turn arise from the phenomenon of hydrogen bonding. The charge displacement between the electronegative oxygen atom and the hydrogen atoms, which may be illustrated as follows:

$$H^{\delta+}\diagup O^{\delta-}-H^{\delta+}-{}^{\delta-}O\diagup{}^{H^{\delta+}}_{\diagdown H^{\delta+}}$$

gives rise to a linear bond between two water molecules with an energy in the range 20–35 kJ mol^{-1}. The interaction energy between water molecules arising

TABLE 1.1. Physical Properties of Some Common Liquids

Property	Water	Methanol	Benzene	n-Hexane
Molecular mass (g $mole^{-1}$)	18.02	32.04	78.12	86.18
Melting point (K)	273	175	279	178
Boiling point (K)	373	338	353	342
Enthalpy of fusion (ΔH_{fus}, kJ mol^{-1})	5.98	2.57	9.9	13.1
Enthalpy of evaporation (ΔH_{evap}, kJ mol^{-1})	40.5	35.2	30.8	28.9
$\Delta H_{fus}/\Delta H_{evap}$	6.8	13.7	3.1	2.2
Heat capacity, C_p (J.mol^{-1} K^{-1})	75.2	2.61	10.6	16.8
Density at 20°C (g cm^{-3})	0.998	0.791	0.879	0.660
Surface tension at 20°C (mNm^{-1})	72.8	22.6	28.9	18.4
Relative permittivity (20°C)	80.1	32.6	2.28	1.89

Data from Weast (1976).

from the hydrogen bond contribution leads to the formation of aggregates of water molecules from dimers to large molecular clusters in which the molecules inside the clusters can form up to four hydrogen bonds with their surrounding neighbors. In this four-point-charge model, due to Bjerrum, the four surrounding molecules were orientated tetrahedrally, but it is known that the tetrahedral is not quite regular. In the solid state at low pressures (ice I), the four-coordinated structure is a very open one with a large amount of empty space. A consequence of this is that ice is very sensitive to pressure and eight other crystalline modifications exist as well as the normal ice I. When ice melts, the structural features of ice I remain to a considerable degree, and it is the partial breakdown of the ice-like structure between 0°C and 4°C that gives rise to the phenomenon of maximum density at 4°C. However, in liquid water molecular clusters still remain above 4°C. The low latent heat of fusion (compared with, e.g., benzene and n-hexane, Table 1.1) and the high latent heat of evaporation reflect the retention of ice-like structure on melting and the breakdown of residual hydrogen bonding on boiling, respectively. The dynamics of the formation and breakdown of the molecular clusters in the liquid state led to them being called "flickering clusters"; their lifetime is of the order of 10^{-10} to 10^{-11} seconds (Frank and Wen, 1957).

Hydrophobicity and hydrophilicity are manifestations of the unique properties of water, and in particular of two essential features, the open hydrogen-bonded structure and the high relative permittivity, respectively. The existence of molecular clusters in the liquid state means that depending on the nature of the solute, the clusters can either be broken down or enhanced. The ease with which many ionic salts dissolve in water, in contrast to their insolubility in organic solvents, relates to the relative ease of separation of the ions due to the high permittivity of water and follows directly from the inverse relationship between the force between ions of opposite charge and the relative permittivity of the medium (Coulomb's Law). In solution, the ions are hydrated by water molecules so that the nearest-neighbor water molecules are largely immobilized by direct ion–dipole interactions. However, surrounding the immobilized water layer is a region of water that differs in structure from bulk water, and for many common salts—e.g., the alkali halides—this region has less ice-like character than bulk water. Such ions are "structure-breaking" and destabilize the flickering clusters, which is manifest in decreases in the partial molal heat capacities of many electrolyte solutions; if there is less "ice-likeness," then less heat is required to raise the temperature of the solution. If, however, we turn to tetraalkylammonium salts, the situation is very different—the apparent molal heat capacities are now larger than expected by up to about 500 J K^{-1} mol^{-1} for, e.g., tetrabutylammonium bromide. This enhanced heat capacity can be regarded as arising from an increase in ice-likeness or a "structure-making" effect due to the inserting of the nonpolar alkyl groups into the water structure. It is the structure-making effect of nonpolar groups in water that is at the root of the hydrophobic effect (see, e.g., Ben Naim, 1980; Tanford, 1973).

The Frank and Wen model of flickering clusters in water and the increasing ice-likeness of water induced by nonpolar solutes was addressed in statistical mechanical terms by Némethy and Scheraga (1962). They considered a distribution of the water molecules over five possible energy levels from the un-hydrogen-bonded state to water molecules in the center of a flickering cluster forming four hydrogen bonds. Molecules at the surface of flickering clusters could form one, two, or three hydrogen bonds. When the energy of the hydrogen bond and the free volume of system are made adjustable parameters, many of the physical properties of liquid water can be accounted for and the structure-forming effect of nonpolar solutes can be quantified as shown in Table 1.2, where the mole fractions of water molecules in the different hydrogen-bonded states are given. For aliphatic and aromatic solutes, the mole fraction of water molecules that are forming four hydrogen bonds is increased by 88% and 65%, respectively, relative to the value in pure water.

The structure-forming effect of nonpolar solutes should not be taken to imply that aliphatic and aromatic groups interact favorably with water; their low solubilities and their adsorption at the aqueous–air interface are manifestations of their antipathy for water, and the intermolecular forces between nonpolar solutes are not particularly strong—in fact fluorocarbons, the most hydrophobic materials, have very weak intermolecular forces. The hydrophobic effect thus relates not to the affinity of nonpolar solutes for themselves (like–like interactions) but to the strong interactions between water molecules, as expressed by Hartley (1936), due largely to hydrogen bonding, which results in nonpolar solutes being "squeezed out" of liquid water because to accommodate them in the water requires reorientation of the water structure with a concomitant decrease in entropy. It is for this reason that the term *hydrophobic bonding*, with the implication of interaction between hydrophobic groups, is to some degree ill chosen to describe the hydrophobic effect. Thermodynamically, the increase in entropy resulting from the breakdown of structured regions around nonpolar residues when they are removed from contact with water molecules makes a major contribution to the lowering of the Gibbs energy of the system and in this sense is the "driving force" behind the hydrophobic effect.

TABLE 1.2. Mole Fractions of Water Molecules in Different Hydrogen-Bonded States at 30°C

Energy State[a]	Pure Water	Aliphatic Solute	Aromatic Solute
0	0.318	0.237	0.256
1	0.237	0.133	0.144
2	0.044	0.183	0.198
3	0.194	0.056	0.060
4	0.207	0.389	0.341

From the model of Némethy and Scheraga (1962).
[a]Number of hydrogen bonds formed.

1.2. TYPES OF AMPHIPATHIC MOLECULES

Amphipathic molecules are ubiquitous in biological systems, and the large surface-to-volume ratio of cells, plasma membranes, cell organelles, and other subcellular structures means that a relatively large proportion of the molecules present are located at interfaces between different bulk phases. A necessary thermodynamic requirement for the existence of a stable interface is that the Gibbs energy of formation of the interface must be positive. If this were not so and the Gibbs energy of formation of the interface were negative, it would grow spontaneously, with the result that one phase would disperse into another. Thus work must be done to extend an interface. The work is done against the interfacial tension and is given by the product of interfacial tension and the change in area. The presence of amphipathic molecules at an interface lowers the interfacial tension and hence reduces the amount of work required to extend it.

In biological systems, numerous processes occur at interfaces, and particularly those associated with membranes such as active, facilitated, and passive transport of substrates and the formation of adenosine triphosphate (ATP) from adenosine diphosphate (ADP) by oxidative phosphorylation and photosynthesis. The stability of biological membranes and the functional systems associated with them depend on the amphipathic nature of their components, specifically the membrane lipids, proteins and glycoproteins. But before discussing the role of amphipathicity in membrane structure, it is appropriate to consider the most important types of amphipathic molecules.

1.2.1. Synthetic Surfactants

Synthetic surfactants are used extensively in the study of biological membranes in order to solubilize membrane components prior to their separation for analysis of composition and/or to get them in a form suitable for manipulation, for example, in the isolation and reconstitution of membrane transport proteins and membrane receptors (Jones et al., 1987). These applications will be discussed in more detail in later chapters. The present discussion is restricted to the types of synthetic surfactants that are extensively used for these purposes.

Surfactants are the active constituents in detergent formulations and generally consist of one or two hydrophobic moieties, typically alkyl chains covalently linked to an ionic or hydrophilic polar headgroup. They can thus be classified in terms of their headgroup as anionic (negatively charged headgroup), cationic (positively charged headgroup), zwitterionic (a headgroup carrying a positive and a negative charge), or nonionic (uncharged but polar headgroup).

Single alkyl chain surfactants are the most frequently used. Figure 1.1 shows the structures of some synthetic surfactants commonly used in membrane studies and their acronyms and/or trade names. The most significant parameter for a surfactant is the concentration at which it forms micelles in solution (see

Anionic Surfactants

$C_{12}H_{25}OSO_3^-$ Na^+ sodium n-dodecylsulphate (SDS)

$C_{12}H_{25}SO_3^-$ Na^+ sodium n-dodecylsulphonate

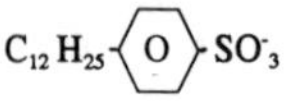

$C_{12}H_{25}$⟨O⟩SO_3^- Na^+ sodium n-dodecylbenzene sulphonate

Cationic Surfactants

$C_{16}H_{31}N^+(CH_3)_3$ Br^- n-hexadecyltrimethylammonium bromide CTAB)

$C_{12}H_{25}N^+(CH_3)_3$ Br^- n-dodecyltrimethylammonium bromide (DTAB)

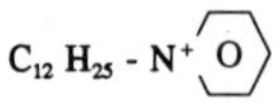

$C_{12}H_{25}-N^+$⟨O⟩ Cl^- n-dodecylpyridinium chloride

Zwitterionic Surfactants

$C_{12}H_{25}-\overset{CH_3}{\underset{CH_3}{\overset{|}{\underset{|}{N^+}}}}-CH_2CH_2CO_2^-$ n-dodecyl-N-betaine

$C_{14}H_{29}-\overset{CH_3}{\underset{CH_3}{\overset{|}{\underset{|}{N^+}}}}-CH_2CH_2CH_2SO_3^-$ N-tetradecyl [N,N-dimethyl-3-ammonia-1-propane sulphonate (Zwittergent 3-14)

Nonionic Surfactants

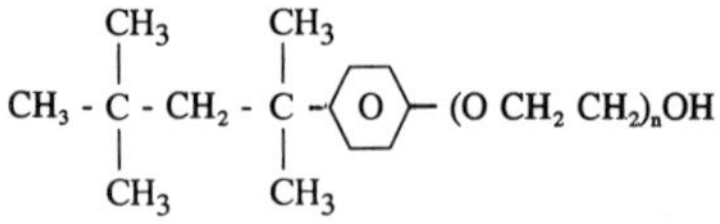

$CH_3-\overset{CH_3}{\underset{CH_3}{\overset{|}{\underset{|}{C}}}}-CH_2-\overset{CH_3}{\underset{CH_3}{\overset{|}{\underset{|}{C}}}}-$⟨O⟩$-(OCH_2CH_2)_nOH$

$C_{12}H_{25}-(OCH_2CH_2)_n\,OH$ Lubrol PX

(polyethylene glycol n-dodecanol)

n = 9 - 10

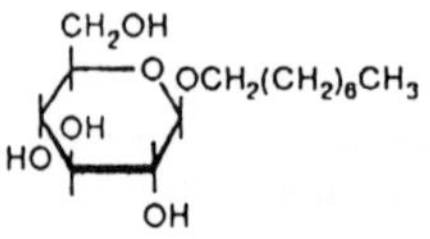

Octyl glucoside (OBG)

(n-octyl-β-D-glucopyranoside)

Triton X-100

(polyethyleneglycol p-t-octylphenol)

n = 9 - 10

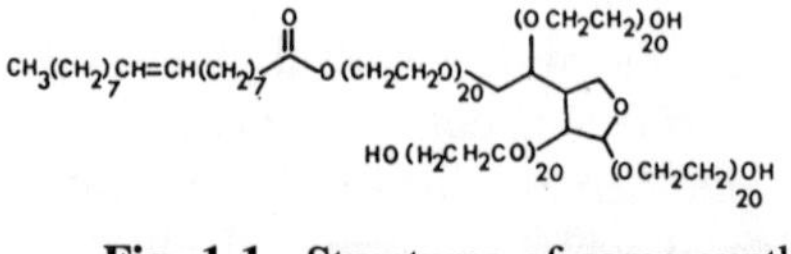

Tween 80

(polyethylene glycol sorbitan monooleate)

Fig. 1.1. Structures of some synthetic surfactants.

Chap. 3), since this governs the maximum concentration of single species (monomers) that can be obtained. It is the activity of the surfactant monomer as distinct from the micelles that is important in many interactions. Depending on the aggregation number of the micelles (i.e., the number of monomers per micelle), the formation of micelles occurs with varying degrees of cooperativity at the so-called critical micelle concentration (cmc). The larger the aggregation number the sharper the cmc. The magnitude of the cmc depends on the hydrophobic-hydrophilic balance of the molecule. Substances with large hydrophobic residues will, in general, more readily form micelles as the solution concentration is increased (i.e., have a lower cmc). The cmc decreases with increasing alkyl chain length and is also affected by the medium. Addition of an inert salt to raise the ionic strength of a solution will lower the cmc of an ionic surfactant.

Many hundreds of surfactants with a wide range of chemical structures have been described although the number that are in common use is small relative to the number of possibilities. The availability, cost, and purity are all important in the choice of a surfactant for a particular application. With regard to purity, it is important to note that the most commonly used surfactant, sodium n-dodecylsulphate (SDS), is difficult to get completely free of its hydrolysis product, n-dodecanol. n-Dodecanol is more surface-active that SDS and consequently competes with the n-dodecylsulphate anion for adsorption sites at interfaces. At the aqueous–air interface, this gives rise to a surface tension-concentration curve with a minimum in the region of the cmc. Below the cmc the alcohol lowers the surface tension more than the anion, but as micelles form the surface tension rises to that of pure anion as the alcohol is solubilized in the micelles. The existence of a minimum in the surface tension–concentration curve of SDS is a stringent test of purity; SDS with a purity of 99.95% gives rise to a minimum of approximately 7 mN m^{-1} (Elworthy and Mysels, 1966). SDS can be purified by continuous diethyl ether extraction but is difficult to maintain completely free of n-dodecanol, as a small degree of hydrolysis rapidly occurs, particularly in acid solutions.

With regard to the use of anionic surfactants, it should also be noted that their divalent and trivalent metal salts have a low solubility, as have the monovalent salts with potassium, cesium, and rubidium ions. For example, the solubilities of potassium and calcium n-dodecylsulphate at 15°C are 3.38 mM and 0.36 mM, respectively. It should also be noted that the solubilities of surfactants do not increase uniformly with temperature but increase sharply at the Krafft point when micelles form. Thus below the Krafft point of approximately 4°C, sodium n-dodecylsulphate crystallizes out of solution, while above 4°C it is readily soluble. These properties are important in the use of surfactants for biochemical processes when choosing appropriate salts for buffers.

The cationic surfactants can be obtained with high purity, although bromides and particularly iodides become contaminated with molecular bromine and iodine, respectively, on storage. The nonionics such as the Tritons, Lubrols, and Tweens with polymeric head groups, depending on the quality, will have a distribution of chain lengths in which the narrower the distribution, the "purer" they can be regarded.

1.2.2. Natural Surfactants

The most commonly known natural surfactants are products of cholesterol metabolism, specifically the bile salts and saponins that are steroid glycosides. The conversion of cholesterol (Fig. 1.2) to the bile acids involves two major structural changes: (1) saturation of the double bond to either bile alcohol 5α (A/B *trans*) or 5β (A/B *cis*) and usually epimerization of the 3-hydroxy group from β to α, and (2) addition of hydroxy groups at the 7 and/or 12 positions and on C_{24} of the side chain, followed by oxidation of the C_{24} alcohol to the acid to give the range of bile acids found in humans shown in Figure 1.3. These are the so-called modern bile acids in contrast to the primitive bile acids of primitive vertebrates (reptiles and amphibians) in which oxidation to the carboxyl group occurs at C_{25}, C_{26}, or C_{27} (Hofmann and Mysels, 1988). The bile "salts" are derived from the corresponding acids by conjugation with glycine ($NH_2CH_2CO_2H$) or taurine ($NH_2CH_2CH_2SO_3H$) by amidation of the carboxyl groups. A significant feature of the bile acids and salts is their stereochemistry. In mammals, the ring structure is almost always 5β (A/B *cis*); this conformation in conjunction with the α conformation of the OH groups results in a molecule with a hydrophobic side and hydrophilic side (Fig. 1.4). Thus in the bile acids and salts, in contrast to the synthetic surfactants, amphipathicity does not arise entirely from the chemistry of the molecules but is imposed by the stereochemical possibilities.

For the purpose of membrane solubilization, the most commonly used bile derivatives are the sodium salts of cholic and deoxycholic acids, which are capable of forming mixed micelles with membrane phospholipids as well as solubilizing proteins without denaturation (see Chap. 3).

The saponins are naturally occurring glycosides, of which the best known are digitonin and ouabain. Both these substances produce physiological effects on the heart, acting as stimulants. Ouabain is a specific inhibitor of sodium/potassium pump in membranes ($Na^+ + K^+ - ATPase$). The structures of these molecules are shown in Figure 1.5, together with a partially synthetically derived surfactant, CHAPS, which is sometimes used in membrane studies.

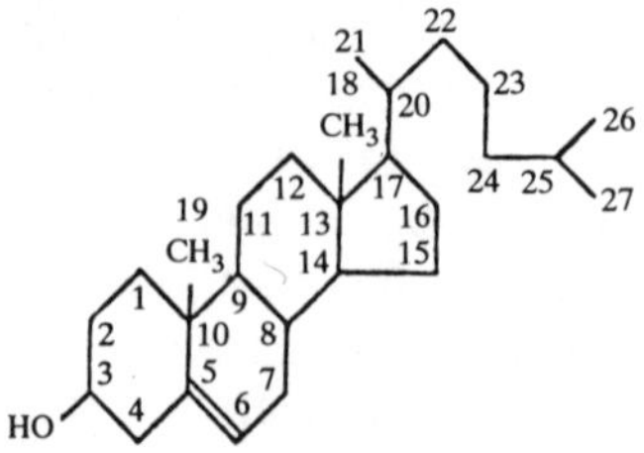

Fig. 1.2. Structure and numbering of carbon atoms in cholesterol.

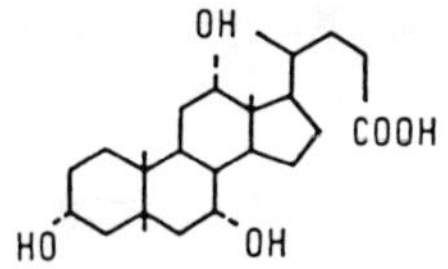

cholic acid

(3α OH, 7α OH, 12α OH)

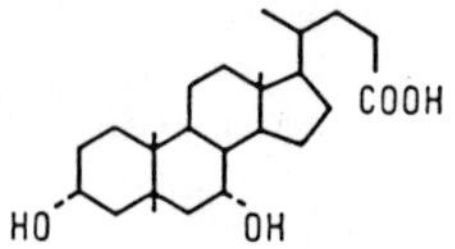

chenodeoxycholic ("chenic") acid

(3α OH, 7α OH)

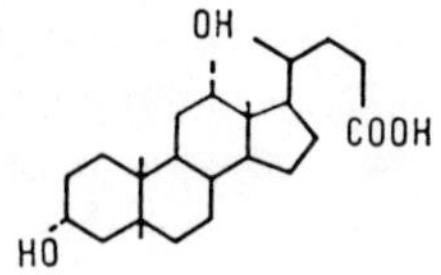

deoxycholic acid

(3α OH, 12α OH)

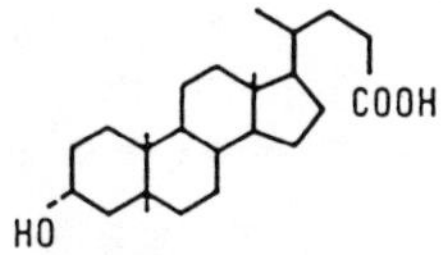

lithocholic acid

(3α OH)

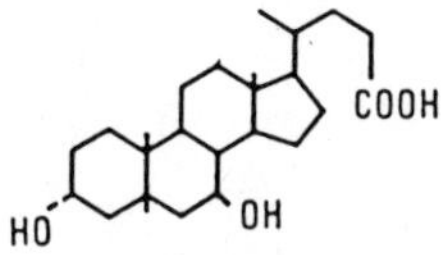

ursodeoxycholic acid

(3α OH, 7β OH).

Fig. 1.3. Chemical structure of the cholic acids found in humans.

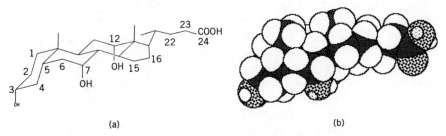

(a)   (b)

Fig. 1.4. Stereochemistry of cholic acid. **a.** Chemical structure. **b.** Molecular model.

Digitonin

(from Digitalis (foxglove))

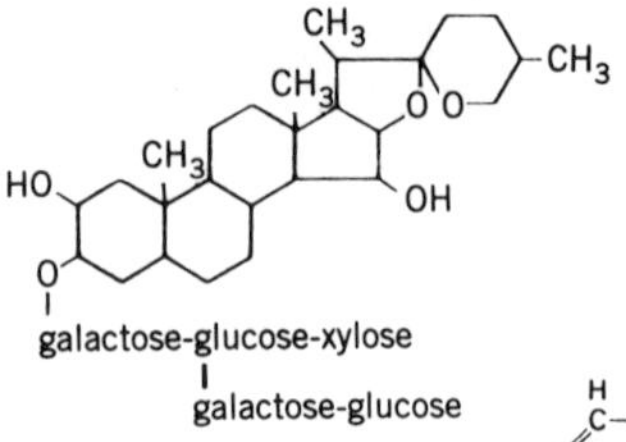

galactose-glucose-xylose
galactose-glucose

Ouabain (G-strophanthin)

(from oubao bark)

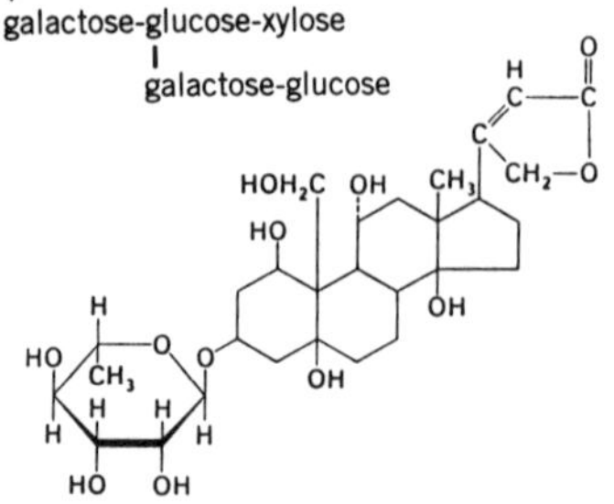

3-[(3-cholamidopropyl)-dimethylammonia]-1-

propane sulphate

CHAPS

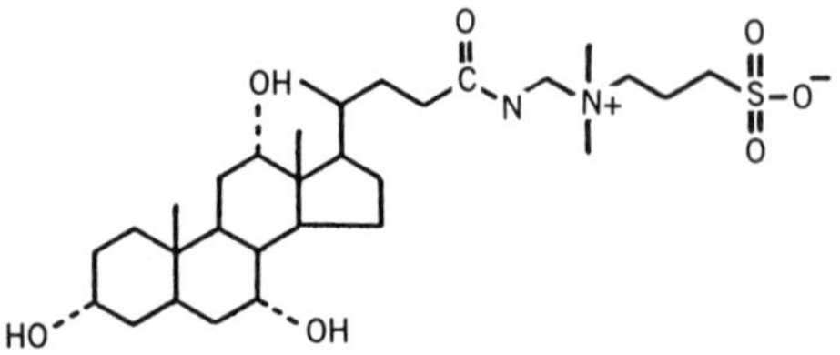

Fig. 1.5. Surfactants structurally related to the bile salts.

1.2.3. Membrane Lipids

The lipids of cell membranes fall into four principal groups: phospholipids, sphingolipids, glycolipids, and sterols. Cholesterol (Fig. 1.2) is the most important member of the last group. Examples of the other three groups are shown in Figure 1.6, and other examples are given in Chapter 7. The phospholipids are derivatives of glycerol, and the sphingolipids (and cerebrosides) are derivatives of sphingosine. The acyl chains (R_1 and R_2 in Fig. 1.6) can be saturated or unsaturated and generally have chain lengths from C_{10} to C_{28}. Some of the commonly occurring acyl chains and "X-group" structures are given in Tables 1.3 and 1.4, respectively. The nomenclature of membrane lipids is complex and has given rise to considerable confusion. In the case of phospholipids, the confusion has arisen because glycerol is a prochiral compound and gives rise to stereoisomers for which different methods of naming have been used, with the result that the same stereoisomers may be referred to by different names. In an attempt to remedy this unsatisfactory situation, Hirschmann (1956) introduced the system of "stereospecific numbering" of the carbon atoms in glycerol, which recognizes the fact that the two primary carbinol groups of glycerol are not identical in their reactions with dissymmetric structures. If the secondary hydroxyl (C-2) group of glycerol is drawn on the left in a Fischer projection of glycerol (Fig. 1.7), the carbon atom above C-2 is called C-1 and

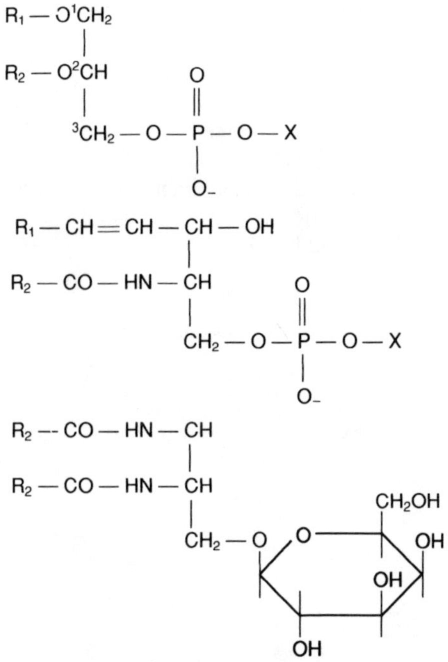

$R_1 - O^1CH_2$
$R_2 - O^2CH$
$^3CH_2 - O - P - O - X$ (with P double-bonded to O above and single-bonded to O_- below)

Phospholipid

$R_1 - CH = CH - CH - OH$
$R_2 - CO - HN - CH$
$CH_2 - O - P - O - X$ (with P double-bonded to O above and single-bonded to O_- below)

Sphingolipid

$R_2 - CO - HN - CH$
$R_2 - CO - HN - CH$
$CH_2 - O$ ... sugar ring with CH_2OH, OH, OH, OH

Glycolipid
(cerebroside)

Fig. 1.6. Membrane lipid structures.

TABLE 1.3. Acyl Chains in Phospholipids

R Group	Systematic Name	Trivial Name	Letter Symbol
$CH_3(CH_2)_{10}CO-$	n-Dodecanoyl	Lauroyl	L
$CH_3(CH_2)_{12}CO-$	n-Tetradecanoyl	Myristoyl	M
$CH_3(CH_2)_{14}CO-$	n-Hexadecanoyl	Palmitoyl	P
$CH_3(CH_2)_{16}CO-$	n-Octadecanoyl	Stearoyl	S
$CH_3(CH_2)_5CH=CH$ $(CH_2)_7CO\text{-}(cis\Delta^9)$		Palmitoleoyl	
$CH_3(CH_2)_7CH=CH$ $(CH_2)_7CO\text{-}(cis\Delta^9)$		Oleoyl	
$CH_3(CH_2)_7CH=CH$ $(CH_2)_7CO\text{-}(trans\Delta^9)$		Elaidoyl	
$CH_3(CH_2)_4CH=CH$ $CH_2CH=CH(CH_2)_7$ $CO\text{-}(cis\Delta^9, \Delta^{12})$		Linoleoyl	
$CH_3CH_2CH=CH$ $CH_2CH=CHCH_2$ $CH=CH(CH_2)_7CO-$ $(cis\Delta^9, \Delta^{12}, \Delta^{15})$		Linoleoyl	
$CH_3(CH_2)_4(CH=CH$ $CH_2)_4(CH_2)_2CO-$ $(cis\Delta^5, \Delta^8, \Delta^{11}, \Delta^{14})$		Arachidonoyl	

TABLE 1.4. Phospholipid Polar Head Groups

X Groups	Name	Letter Symbol
$-H$	Phosphatidic acid	PA
$-CH_2CH_2N^+(CH_3)_3$	Phosphatidylcholine (lecithin)	PC
$-CH_2CH_2NH_3^+$	Phosphatidylethanolamine (cephalin)	PE

$-CH_2-CH-CO_2.$
      $\mid$
     NH_3^+
	Phosphatidylserine	PS

(inositol ring structure)	Phosphatidylinositol	PI

the one below C-3. Thus the C-1 and C-3 carbinol groups are distinguished and used to describe the stereochemistry of the derivatives. The use of this stereospecific numbering is denoted by the prefix *sn* before the stem name of the compound. Some typical examples of the use of this system of nomenclature are

1,2-dimyristoyl-*sn*-glycero-3-phosphocholine (dimyristoylphosphatidylcholine, DMPC)

1,2-dipalmitoyl-*sn*-glycero-3-phosphoserine (dipalmitoylphosphatidylserine, DPPS)

1-myristoyl-2-palmitoyl-*sn*-glycero-3-phosphoethanolamine (1-myristoyl-2-palmitoylphosphatidylethanolamine, MPPE)

Here the symbol P denotes the phosphatidyl group. Such one-letter nomenclature is in common use, although the symbol Ptd has been recommended (IUPAC-IUB, 1967, 1977).

The phosphorylated derivative of glycerol, *sn*-glycerol-3-phosphate, has the

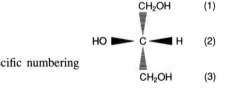

Fig. 1.7. Fischer projection and stereospecific numbering of glyercol.

L-absolute configuration (D/L, Fischer nomenclature) or the S (sinister) absolute configuration (R/S [rectus/sinister], Cahn, Ingold, and Prelog nomenclature). The value of stereospecific numbering is immediately appreciated when it is realized that by using the D/L or R/S nomenclature sn-glycerol-3-phosphate could be called L or S (glycerol 3-phosphate) and D or R (glycerol 1-phosphate), which is clearly a source of potential confusion. When the compound is a racemic mixture, the prefix rac is used in place of sn, and if the configuration is unknown the prefix X is used. A further point of nomenclature for phospholipids is the term $lyso$-, used to denote a phospholipid in which one of the two acyl groups has been hydrolytically removed; the location of the cleaved chain can be denoted by 1-lyso or 2-lyso. The lyso prefix originates from the fact that lyso compounds destabilize the lipid bilayer of cell membranes and lead to cell lysis.

1.2.4. Amphipathic Nature of Proteins

The 20 amino acids that are used in the formation of polypeptide chains in proteins can be broadly defined as having polar or nonpolar side chains. The polar side chains such as glutamic and aspartic acids will be ionized at physiological pH, conferring negative charge on a protein while those of lysine, arginine, and histidine ionize to confer positive charge. Fisher (1964) showed that in general the larger a globular protein the lower its overall polarity, in that larger globular proteins can accommodate a greater proportion of nonpolar residues in their interior than small proteins, which have relatively more of their amino acid residues exposed to solvent. The amphipathic properties of proteins arise directly from the properties of the amino acid residues, and they play a major role in the overall conformation and stability of the native state (Bigelow, 1967). There are numerous interactions in the native conformation of both globular and fibrous proteins that contribute to their stability. These include hydrogen bonding—both intramolecular hydrogen bonding between the NH and CO groups of the polypeptide chain every fourth residue along the backbone of the α-helix and intermolecular hydrogen bonding between β-pleated sheets—as well as salt linkages between positively and negatively charged amino acid side chains, dipole–dipole, ion–dipole, and dispersion forces. However, as was highlighted by Kauzmann (1959), hydrophobic interactions between nonpolar residues play a significant part in stabilizing native conformations. Table 1.5 shows an attempt made by Finney (1982) to estimate the various contributions to the overall stability of the native state of proteins. As the range of figures demonstrates, it is difficult to draw up a balance sheet of this type with any certainty even for proteins such as lysozyme or ribonuclease where the structures are known from X-ray studies. Furthermore, it is not possible to obtain reliable estimates for some of the contributions such as charge–charge interactions on the protein surface (probably very small), changes in van der Waals interactions (although partly accounted for in the hydrophobic interaction), or contributions from vibrational entropy changes. However, a number of significant points can be made from the estimates in Table 1.5, the most important being that many of the contributions are large and differ in sign. For example, the hydrophobic interactions and associated gain in configurational entropy from release of water molecules amount to approximately $-1,000$ kJ mole^{-1}, whereas the distortion of H-bonds in the native state and the loss of configurational entropy of the polypeptide chain each contributes up to $+800$ KJ mol^{-1}. Thus the native state is maintained by a delicate balance between apposing high-energy contributions, and hence any uncertainties in the estimates of these high-energy contributions have a profound effect on the net energy balance. The estimates in Table 1.5 show, however, that without the hydrophobic contribution and the accompanying entropy change from water "destructuring" the net Gibbs energy change on folding would be positive and hence folding would be unfavorable.

In principle, if the Gibbs energies of transfer of every amino acid residue from the aqueous phase to its specific location in the interior of a protein in its native state were known, then the stabilization energy of the native conformation could be calculated from the primary sequence. This seems such a desirable objective that much effort has been directed, and continues to be directed, toward obtaining reliable estimates for energies of transfer. The earliest attempts to make such estimates were based on the differences in solubility of each amino acid between water and an organic solvent that was considered to most appropriately represent the interior of the globular protein and the corresponding differences in solubility for the simplest amino acid, glycine.

TABLE 1.5. Contribution to the Stability of the Small Globular Proteins Lysozyme (Relative Molecular Mass 14,306) and Ribonuclease (Relative Molecular Mass 13,682)[a]

Process	Estimated Contribution (kJ mole^{-1})
Changes in ionization of charged amino acids, i.e., ΔpKa	-20
Changes in hydrogen bonding	
1. water–polar group $\rightarrow$ polar–polar group	-300
2. distortion of H-bonds	$+400 \rightarrow +800$
3. unsaturation of H-bonds in the native state	300
Hydrophobic interactions	-550
Loss of configurational entropy	$+300 \rightarrow +800$
Gain of configurational entropy from release of water molecules	-500
Net Gibbs energy change on folding	$+530 \rightarrow -370$
	Mean $+ 80$
Experimental Gibbs energy of folding	-61 (Lysozyme)
	-59 (Ribonuclease)

[a]The estimates relate to the transition from the unfolded (denatured) to the folded (nature) state. Adapted from Finney (1982).

The organic solvents chosen for this purpose were, for example, methanol, ethanol, and also n-octanol. The principle of the solubility method is that for saturated solutions of an amino acid aa in water (w) and in the chosen organic solvent (o), the chemical potentials ($\mu_i^w = \mu_i^o$) are equal, thus in water

$$\mu_{aa}^w = \mu_{o\,aa}^w + RT \ln x_{aa}^w \tag{1.1}$$

in the organic solvent

$$\mu_{aa}^o = \mu_{o\,aa}^o + RT \ln x_{aa}^o \tag{1.2}$$

where $\mu_{o\,aa}^w$ and $\mu_{o\,aa}^o$ are the standard chemical potentials of aa in water and the organic solvent and x_{aa}^w and x_{aa}^o the corresponding mole fractions at saturation.

Thus since $\mu_{aa}^w = \mu_{aa}^o$ at saturation, equating (1.1) and (1.2) gives

$$\mu_{o\,aa}^o - \mu_{o\,aa}^w = \delta G_{tr,\,aa} = RT \ln \left(\frac{x_{aa}^w}{x_{aa}^o}\right) \tag{1.3}$$

where $\delta G_{tr,\,aa}$ is the Gibbs energy of transfer of aa from water to the organic solvent. For glycine (gly)

$$\mu_{o\,gly}^o - \mu_{o\,gly}^w = \delta G_{tr,\,gly} = RT \ln \left(\frac{x_{gly}^w}{x_{gly}^o}\right) \tag{1.4}$$

Hence

$$\Delta G_{trs,\,aa} = \delta G_{tr,\,aa} - \delta G_{tr,\,gly} = RT \ln \left(\frac{x_{aa}^w}{x_{aa}^o}\right) - RT \ln \left(\frac{x_{gly}^w}{x_{gly}^o}\right) \tag{1.5}$$

where $\Delta G_{tr,\,aa}$ represents the Gibbs energy of transfer of the amino acid side chain of aa from water to the organic solvent, since by subtracting glycine the contributions of the amino, carboxyl, and the α-carbon atom (plus one hydrogen) moieties are eliminated. Although this procedure appears to be a simple means of estimating the hydrophobicities of amino acid side chains, since the more negative the value of $\Delta G_{tr,\,aa}$ the stronger the hydrophobic interaction, there are several problems in establishing the best set of values to use and various refinements to this simple analysis have been introduced and numerous "polarity scales" have been derived (Guy, 1985). Clearly the choice of organic solvent is important, but also the fact that in proteins, depending on their location in the structure, amino acid residues are exposed to water to various extents. A simple classification into exposed and buried residues, as the above analysis implies, is an oversimplification. In an attempt to take account of the location of amino acid residues in proteins, Guy (1985) developed a layer analysis in which residue distributions were analyzed as a function of distance from the protein surface. Using a statistical analysis of a group of 19 proteins

TABLE 1.6. Transfer Gibbs Energies for Amino Acid Residues (Polarity Scales)[a]

Amino Acid	$\Delta G_{tr,aa}$ (kJ mol^{-1})		
	1	2	3
Phe	−7.03	−11.1	−15.5
Cys	−5.69	—	− 8.37
Ile	−5.48	−12.4	−13.0
Met	−5.31	− 5.44	−14.2
Leu	−5.06	−10.1	−11.7
Val	−4.56	− 7.03	−10.9
Trp	−3.68	−12.6	− 7.95
His	−2.05	− 2.80[b]	+12.6
Tyr	−1.38	−10.6	+ 2.93
Ala	+0.25	− 3.05	− 6.69
Thr	+1.13	− 1.84	− 5.02
Gly	+1.72	—	− 4.18
Asn	+2.01	+ 0.04	+20.1
Ser	+2.10	− 0.17	− 2.51
Pro	+2.93	—	+ 0.837
Gln	+3.05	+ 0.42	+17.2
Glu	+3.22	− 2.30[b]	+34.3
Asp	+3.35	− 2.26[b]	+38.5
Arg	+3.51	− 3.05[b]	+51.5
Lys	+4.94	− 6.28[b]	+36.8

[a]Scale 1 (Guy [1985] layer theory, water → protein interior). Scale 2 (Nozaki and Tanford [1971], water → ethanol). Scale 3 (Engelman, Steitz and Goldman [1986], water → α helix in a membrane).
[b]Values for the unionized state.

carried out by Prabhakaran and Ponnuswany (1980), distribution curves for the amino acids were obtained, representing the energies required to move a residue from a reference position to a given layer located at a particular distance from the surface of the protein. From such distribution curves and the Gibbs energies required to transfer residues from water to a completely buried position within the protein structure, a polarity scale was developed that is probably the most reasonable scale to use for the Gibbs energies of transfer of a given amino acid residue from the aqueous phase to the interior of a protein. Table 1.6 (column 2) shows the polarity scale based on layer theory in comparison with a previous scale (column 3) based on Gibbs energies of transfer between water and ethanol. It should be stressed that the figures in such polarity scales are best estimates and are probably only accurate to ±30%. The scale based on layer analysis is, however, a marked advance over the previous scales, and, while the absolute values of the transfer energies are not particularly accurate, for many purposes it is the relative values of the amino acids that are important. In this respect, the order of the hydrophobicities of the amino acid are reasonably well defined, phenylalanine being the most hydrophobic and lysine in the charged state the least hydrophobic.

1.3. THE ROLE OF AMPHIPATHIC MOLECULES IN BIOMEMBRANES

Membranes have been shown to consist of a bilayer arrangement of lipids in association with proteins and glycoproteins, as depicted in the classical model of Singer and Nicolson (1972). The mass ratio of protein to lipid in membranes ranges from approximately 0.25 in the myelin membrane around the axon of the spinal chord to approximately 3.6 in the inner membrane of mitochondria, the site of oxidative phosphorylation. The ratio of protein to lipid reflects the nature and extent of the biochemical functions that a given membrane has to perform. The ratio of cholesterol to lipid in mammalian cell membranes also varies over a relatively wide range, from approximately 1 in myelin and erythrocyte plasma membranes to as low as 0.2 in the inner mitochondrial membrane. The plasma membranes of bacteria contain no cholesterol, and hence the fluidity of the bilayer is governed only by the length and unsaturation of the alkyl lipid chains, whereas in mammalian membranes the cholesterol level plays a role in keeping the membrane fluid by interaction with other lipids.

Membrane proteins can be broadly divided into those that penetrate to varying degrees into or through the bilayer matrix—the so-called integral or intrinsic membrane proteins or glycoproteins—and those that are adsorbed on the membrane surface—the so-called peripheral or extrinsic proteins or glycoproteins (see Chapter 7).

The hydrophobic interaction between membrane proteins and the alkyl lipid chains is central to both the structure and function of the membrane. In general, the lipid bilayer is approximately 7.5 nm thick, and to cross the hydrophobic barrier presented by the alkyl lipid chains (~ 3 nm) requires a sequence of approximately 20 hydrophobic amino acid residues in the α-helical conformation. It is generally found from circular dichroism studies on transmembrane proteins that the transmembrane segments of the proteins are in the α-helical conformation, which results in the hydrophobic side chains of the amino acid residues being exposed on the helical surface to facilitate hydrophobic interaction with the lipid chains. Table 1.7 shows the subunit composition and secondary structure of some integral membrane proteins and the numbers of α-helical membrane spanning regions which are believed to be present in the transmembrane regions of the proteins. For many of the membrane transporters and receptors, the primary sequences have been obtained from the gene sequence of the cDNA and the transmembrane α-helical sequences have been identified with reference to a hydropathy plot constructed using a polarity scale for the amino acids in order to identify amino acid sequences that are sufficiently hydrophobic and sufficiently long to suggest that they could exist in a transmembrane helix.

TABLE 1.7. Integral Membrane Protein Structure

Protein	Relative Molecular Mass	Subunit Composition	α-Helical Membrane Spanning Regions
Erythrocyte anion transporter	90,000	Single chain	14
Erythrocyte glucose transporter	55,000	Single chain	12
Na^+-K^+ ATPase (Na^+-K^+ pump)	304,000	$\alpha_2\beta_2$	7 (α chain) 4 (β chain)
Ca^{2+}-ATPase (sarco-plasmic reticulum, Ca^{2+} pump)	110,000	Single chain	10
Bacteriorhodopsin (Halobacterium halobium, H^+ pump)	26,000	Single chain	7
Insulin receptor (adipocytes)	460,000	$\alpha_2\beta_2$	2 (β chain)
β-Adrenergic receptor	64,000	Single chain	7
Acetylcholine receptor	268,000	$\alpha_2\beta\gamma\delta$	~40
GABA(γ-aminobutyric acid) receptor (brain)	220,000	$\alpha_2\beta_2$	4 (α chain) 4 (β chain)
K-receptor (neuropeptide receptor)	43,066	Single chain	8
Rhodopsin (photoreceptor of retinal rods)	40,000	Single chain	7
Glycophorin A (erythrocyte sialoglycoproteins)	30,000	Single chain	1

Data from: Stryer (1988); for the glucose transporter, Mueckler et al. (1985); for Ca^{2+}-ATPase, Anderson and Vilsen (1990); for the K-receptor, Masu et al. (1987).

The problems of establishing a suitable polarity scale appropriate for the transfer of amino acid residues from the aqueous environment to an α-helix embedded in a hydrophobic environment are similar to those discussed above for estimating the Gibbs energies of transfer from an aqueous medium to the interior of the native state of a protein. Numerous scales have been derived based on solubility measurements, vapor pressures of side-chain analogues, and the analysis of side-chain distributions in soluble proteins.

Table 1.6 (column 4) shows the scale devised by Engelman et al. (1986), which was developed by consideration of the nonpolar properties of the amino acids as they exist in a helix using a semitheoretical method in combination with experimental values for the polar and nonpolar characteristics of amino acid side chains. The polarity scales for the transfer from water to protein interior and water to α-helix in a membrane differ significantly, as these two processes are not in general analogous. After establishing a polarity scale, the procedure for construction of a hydropathy plot consists of calculating the Gibbs energy of transfer of a sequence of 20 amino acid residues from the aqueous phase to a hypothetical α-helix, starting with the first residue in the protein sequence and repeating the calculation from residues 2–21, 3–22, 4–23, etc. throughout the sequence. The Gibbs energy of transfer per amino acid (ΔG_{trs}/20) is plotted as a function of the first amino acid in the 20-amino acid "win-

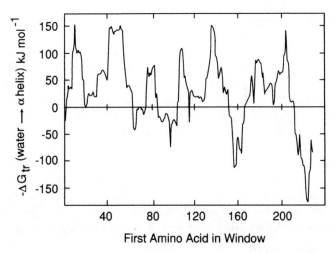

Fig. 1.8. Hydropathy plot for bacteriorhodopsin. From Engelman et al. (1986), with permission of the publisher.

dow'' to form the hydropathy plot. Figure 1.8 shows a hydropathy plot for bacteriorhodopsin which shows seven well-defined maxima which are thought to correspond to the seven membrane-spanning regions.

There are a number of problems associated with both the construction and interpretation of hydropathy plots even given a reliable polarity scale. The choice of the amino acid window appropriate to a given membrane protein will depend on both the thickness of the membrane bilayer, which will relate to the particular lipid composition and the cholesterol content, and to the orientation of the helix with respect to the plane of the bilayer. Membranes can vary in thickness by as much as a factor of two, although the hydrophobic barrier is generally of the order of 3 nm thick, which corresponds to a 20–21 residue window. After establishing an appropriate window, the next concern is how significant a peak in the hydropathy plot is, which depends on the energy criterion used. From studies on membrane proteins that are definitely known to span the membrane, it is generally considered that a peak of approximately 80 KJ mol^{-1} correlates with the stable insertion and anchoring of a protein in a membrane. However, it should be noted that peaks of this magnitude can sometimes be found in hydropathy plots of nonmembrane proteins, so that some caution is required in the assignment of membrane-spanning regions to sequences, and other types of experimental evidence are required to confirm predictions based on hydropathy plots. Despite the approximations used in the construction of hydropathy plots and the limitations in their interpretation, they nevertheless clearly demonstrate the importance of the hydrophobic interaction in anchoring functional membrane proteins to the bilayer matrix. It should be noted, however, that anchoring by α-helical domains is not the only method of anchoring proteins and glycoproteins to membranes that is dependent on the hydrophobic interaction. Some membrane-associated proteins are anchored by covalent modifications, which include addition of myristic acid through an amide bond to the α-amino group of amino-terminal glycine, addition of palmitic acid—usually linked through a thioester to cysteine—or addition of a complex phospholipid at the C-terminus (Sefton and Buss, 1987). The addition of these hydrophobic anchors is not uncommon, and a relatively large number of functionally important membrane-bound proteins of eukaryotic cells—including, for example, acetylcholine esterase, 5′-nucleotidase, alkaline phosphatase (all having covalently linked phospholipid), NADH-cytochrome b_5 reductase (myristylated), rhodopsin, and ankyrin (palmitylated)—are anchored to membranes by these structural features. It is not clear why this method of anchoring has evolved, although a possible advantage of this method is that it could possibly confer greater lateral mobility on such surface-bound proteins than would occur if they were anchored by larger α-helical domains.

1.4. THE ROLE OF AMPHIPATHIC STRUCTURES IN PROTEIN TARGETING

It is well established that when mRNA for membrane and secretory proteins is translated *in vitro*, the proteins produced contain an extra sequence of 15–30 amino acid residues on their N-termini that is absent when the proteins are isolated from the cell. These extra sequences are targeting signals which *in vivo* are required to ensure that a given membrane or secretory protein reaches its required destination. After targeting of the nascent chain to the membrane, the signal sequence is cleaved, leaving the mature protein. Signal sequences are required for the selective targeting of nascent protein chains to various sites, including the endoplasmic reticulum in eukaryotes, or to the cytoplasmic membrane in prokaryotes. Despite extensive study, the details of the mechanism of the process by which the signal sequences facilitate translocation across the cytoplasmic membrane or the endoplasmic reticulum are not clearly understood (Gierasch, 1989).

The structures of signal sequences appear to follow a general pattern that is the same for both eukaryotes and prokaryotes and consists of three elements together with a targeting domain that varies with the target site (in Chapter 9). The primary protein targeting signals are for the endoplasmic reticulum, the mitochondrial matrix, the chloroplast, the peroxisomes, and the cell nucleus (Von Heijne (1990). Figure 1.9 shows the overall structure of a composite targeting signal, which consists of a targeting domain and a signal peptide with three regions, c, h, and n. Progressing from the signal peptidase cleavage site, the c region usually consists of 5 to 7 amino acid residues, which are of high polarity although not generally charged. The h region is the most characteristic feature of the signal peptide and consists of 10 ± 3 amino acids rich in Leu, Ala, Met, Val, Ile, Phe, and Trp but may contain Pro, Gly, Ser, or Thr residues. The overall hydrophobicity of the h region is an essential requirement.

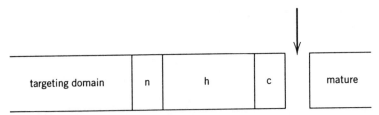

Fig. 1.9. Composite targeting signal showing the targeting domain followed by a signal peptide with n, h, and c regions and the cleavage site (↓) between the signal and the mature protein. Adapted from Von Heijne (1990).

The n region is highly variable in length and composition, but always carries a net positive charge. Signal sequences display a surprising lack of primary sequence homology even among closely related proteins. After the signal sequence is synthesized and the nascent polypeptide chain emerging from the ribosome is approximately 70–80 residues long, it binds to a signal recognition particle (SRP). Current evidence suggests that only one SRP is present in a particular organism, so that different signal sequences must be recognized by the same SRP. The next step in the mechanism is unclear. There is evidence that the endoplasmic reticulum contains a signal sequence receptor (SSR) or that the nascent chain binds to phospholipids of the membrane, and that this initial binding to either an SSR or to the bilayer is followed by interaction with some protein or complex that facilitates translocation. Whatever the details of the process are, it is clear that the signal sequence is required to carry out a number of roles, including binding to the SRP, binding to the membrane to be translocated, facilitation of translocation, and recognition by the signal peptidase. The absence of specific patterns of amino acid sequence in signals does not seem consistent with the presence of specific protein–protein interactions, although the n and h regions of the signal peptide, being positively charged and hydrophobic, respectively, appear to be ideally suited for interaction with the negative surface and hydrophobic interior of membranes.

The study of isolated signal peptides has shown that although they adopt either a random coil or β-pleated sheet conformation in aqueous solution, they have a conformational preference for forming an α-helix in nonpolar solvents or in the hydrophobic interior of a micelle. The ability to adopt the α-helical conformation in a nonpolar environment thus appears to be a property of signal sequences, although this alone is not sufficient for a given sequence to act as a signal. However, signal sequences are an interesting example of amphipathicity playing an important role in a specific biochemical process.

REFERENCES AND FURTHER READING

Andersen, J.P. and Vilsen, B. (1990) Primary ion pumps. Curr. Opin. Cell Biol. 2, 722–730.

Ben Naim, A. (1980) Hydrophobic Interactions. Plenum, New York.

Bigelow, C.C. (1967) On the average hydrophobicity of proteins and the relationship between it and protein structure. J. Theor. Biol. *16*, 187–211.

Elworthy, P.H. and Mysels, K.J. (1966) The surface tension of sodium dodecylsulphate solutions and the phase separation model of micelle formation. J. Coll. Int. Sci. *21*, 331–347.

Engelman, D.M., Steitz, T.A. and Goldman, A. (1986) Identifying non-polar trans-bilayer helices in amino acid sequences of membrane proteins. Annu. Rev. Biophys. Chem. *15*, 321–353.

Finney, J.L. (1982) in Biophysics of Water. F. Franks and S.F. Mathias, Eds. J. Wiley and Sons, Chichester, pp. 55–58.

Fisher, H.F. (1964) A limiting law relating the size and shape of protein molecules to their composition. Proc. Natl. Acad. Sci. USA *51*, 1285–1291.

Frank, H.S. and Wen, W.Y. (1957) Ion–solvent interactions: Structural aspects of ion-solvent interaction in aqueous solutions: A suggested picture of water structure. Discuss. Faraday Soc. no. 24, 133–140.

Franks, F. (1983) Water. Royal Society of Chemistry, London.

Gierasch, L.M. (1989). Signal sequences. Biochemistry *28*, 923–930.

Guy, H.R. (1985) Amino acid side-chain partition energies and distribution of residues in soluble proteins. Biophys. J. *47*, 61–70.

Hartley, G.S. (1936) Aqueous Solutions of Paraffin Chain Salts. Hermann and Cie., Paris.

Helenius, A. and Simons, K. (1975) Solubilization of membranes by detergents. Biochim. Biophys. Acta *415*, 29–79.

Hirschmann, H. (1956) The structural basis for the differentiation of identical groups in asymmetric reactions, in Essays in Biochemistry. S. Graff, Ed. John Wiley & Sons, Chichester, pp. 156–174.

Hofmann, A.F. and Mysels, K.J. (1988) Bile salts as biological surfactants. Colloids Surf. *30*, 145–173.

Houslay, M.D. and Stanley, K.K. (1982) Dynamics of Biological Membranes. John Wiley and Sons, Chichester.

IUPAC-IUB Commission on Biochemical Nomenclature. The nomenclature of lipids (1967) J. Biol. Chem. *242*, 4845–4849; and (1977) Eur. J. Biochem. *79*, 11–21.

Jones, O.T., Earnest, J.P. and McNamee, M.E. (1987) in Biological Membranes. J.B.C. Findlay and W.H. Evans, Eds. IRL, Oxford, pp. 139–177.

Kauzmann, W. (1959) Some factors in the interpretation of protein denaturation. Adv. Protein Chem. *24*, 1–63.

Lichtenberg, D., Robson, R.J. and Dennis, E.A. (1983) Solubilization of phospholipids by detergents. Biochim. Biophys. Acta *737*, 285–304.

Masu, Y., Nakayama, K., Tamaki, M., Haruda, Y., Kuno, M. and Nakanishi, S. (1987) cDNA cloning of bovine substance-K receptor through oocyte expression system. Nature *329*, 836–840.

Muecklen, M., Caruso, C., Baldwin, S.A., Pancio, M., Biench, I., Morris, H.R., Allard, W.J., Leinhard, G.E. and Lodish, H.F. (1985) Sequence and structure of a human glucose transporter. Science *229*, 941–945.

Némethy, G. and Scheraga, H.A. (1962) Structure of water and hydrophobic bonding in proteins. Part 1: A model for the thermodynamic properties of liquid water. Part 2: Model for the thermodynamic properties of aqueous solutions of hydrocarbons. J. Chem. Phys. *36*, 3382–3400; 3401–3417.

Nozaki, Y. and Tanford, C. (1971) The solubility of amino acids and two glycine peptides in aqueous ethanol and dioxane solutions. J. Biol. Chem. *246*, 2211–2217.

Prabhakaran, M. and Ponnuswany, P.K. (1980) Spatial assignment of amino acid residues in globular proteins: An approach from information theory. J. Theor. Biol. *87*, 623–637.

Sefton, B.M. and Buss, J.E. (1987) The covalent modification of eukaryotic proteins with lipids. J. Cell Biol. *104*, 1449–1453.

Singer, S.J. and Nicolson, G.L. (1972) The fluid mosaic model of the structure of cell membranes. Science *175*, 720–731.

Stryer, L. (1988) Biochemistry. 3rd ed. W.H. Freeman & Co., San Francisco.

Szent-Györgyi, A. (1957) Bioenergetics. Academic Press, New York.

Tanford, C. (1973). The Hydrophobic Effect: Formation of Micelles and Biological Membranes. John Wiley and Sons, New York.

Von Heijne, G. (1990) Protein targeting signals. Curr. Opinion in Cell Biol. *2*, 604–608.

Weast, R.C., Ed. (1976) Handbook of Chemistry and Physics. 57th ed. CRC Press, Cleveland, Ohio.

CHAPTER 2

MONOLAYERS

2.1. THE FORMATION OF MONOLAYERS

The concept that amphipathic molecules, which are almost insoluble in aqueous media, can form monomolecular layers at the aqueous–air interface was established toward the end of the nineteenth century as a consequence of the pioneering work of Fraulein Pockels and Lord Rayleigh. A necessary structural requirement for the formation of a layer one molecule thick at the aqueous–air interface is the combination of a relatively large hydrophobic group with a polar or charge headgroup. The hydrophilic nature of the polar or charged headgroup anchors the molecule to the aqueous interface and allows the hydrophobic group to protrude toward the gaseous (air) phase. When the molecule as a whole is not significantly soluble in the aqueous phase, the monolayer is generally formed by spreading from an organic solution. Such monolayers are called "spread films" or "insoluble monolayers," in contrast to monolayers formed at the aqueous–air interface from water-soluble amphiphiles, which are termed "Gibbs monolayers," since here the relationship between surface tension and solute concentration is governed by the Gibbs adsorption isotherm. The interfacial surface density of molecules in insoluble monolayers is controlled by the amount of amphiphile placed on the interface and the area available to it, whereas in the case of Gibbs monolayers there is an equilibrium between the monolayer and the solute in the aqueous (substrate) phase. Thus the interfacial surface density of molecules in a Gibbs monolayer will vary with the solute concentration in the aqueous phase. There have been numerous texts (Adam [1941] 1968; Gaines, 1966; Davies and Rideal, 1963; Aveyard and Haydon, 1973; MacRitchie, 1990) that cover early work on monolayers and the detailed experimental aspects of the field. The objective of this chapter is to consider the part played by monolayer studies in the understanding of biological interactions and processes in membranes, with particular regard to more recent selected studies of particular significance.

In the context of membrane studies, the stimulus to use monolayers as a means of understanding the behavior of processes in biological membranes stems from the classical work of Gorter and Grendel in 1925. These workers attempted to extract the lipid from the membranes of a number of erythrocytes (red blood cells) of different species with acetone and spread them at the aqueous–air interface on a Langmuir trough (see sec. 2.2), the objective being to determine the limiting area covered by the close-packed monolayer of the lipids. They found that the area covered by a monolayer of the erythrocyte plasma membrane lipid was approximately twice the surface area of the erythrocyte, and so deduced that the plasma membrane was a bilayer resembling the orientation of fatty acids in crystals, as previously suggested by Bragg. While the significance of this observation in the field of membrane studies can hardly be overestimated, it is interesting to note that the details of the experimental procedures that led to the deduction of a bilayer structure were far from correct. Specifically, the surface areas of the erythrocytes were not then accurately known; the method of lipid extraction was not sufficiently quantitative; they assumed that the surface pressure in the plasma membrane was zero (see sec. 2.8); and the area taken up by the integral membrane proteins was neglected. Despite the shortcomings of the experiments in the light of current knowledge, the results they obtained are not in doubt; although a later study by Dervichian and Macheboeuf (1938) suggested the plasma membrane is a lipid monolayer, this result was largely ignored. In 1966 a reinvestigation of the problem by Bar et al. showed that at low surface pressures, the ratio of monolayer area to erythrocyte surface area was 2 : 1, and that only at the monolayer collapse pressure (see sec. 2.2) did the ratio become 1 : 1.

In that the lipid bilayer is two monolayers placed back to back, with the charged or polar residues protruding into the aqueous phase, thus satisfying the amphipathicity of the lipid molecules, the study of the properties and interactions with and within monolayers is of direct relevance to the bilayer and hence to biological membranes. The value of the monolayer technique relates to the facts that their composition can be precisely controlled and they can be manipulated through variations in both surface parameters and aqueous substrate composition, thus enabling experiments to be set up to answer specific questions concerning individual membrane components without the complications arising from multiple interactions with other components that are present in cell membranes.

2.2. TECHNIQUES USED TO STUDY MONOLAYERS AT LIQUID INTERFACES

The oldest and still the most commonly used technique for the study of monolayers at the aqueous–air interface is the Langmuir-Adam surface balance, often

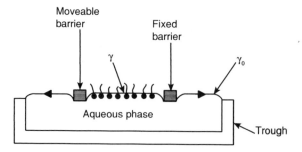

Fig. 2.1. The principle of the Langmuir-Adam film balance.

called the Langmuir trough. Figure 2.1 shows the principle features of the Langmuir trough. It consists of a rectangular trough made from a hydrophobic material such as Teflon (polytetrafluoroethylene) that is filled with the aqueous subphase (substrate). Two barriers cross the trough, a movable barrier and a fixed barrier or float connected to the sides of the trough by floating hydrophobic threads. The area between the barriers can be varied by movement of the movable barrier. The monolayer is spread from an organic solvent when a known amount of a lipid in an organic solvent is placed between the barriers and the organic solvent is allowed to evaporate. The surface tension (γ) between the barriers is lowered by the monolayer relative to the clean aqueous surface of surface tension (γ_0), and as a consequence there is a "surface pressure" (π) on the barriers in the direction of the clean aqueous surfaces. The surface pressure is defined by the equation

$$\pi = \gamma_0 - \gamma \tag{2.1}$$

The surface pressure can be measured either by measuring the force per unit length on the fixed barrier by mechanically attaching it to a calibrated torsion wire, or by using the Wilhelmy plate technique to measure γ of the monolayer. It follows that since the surface pressure arises from the difference in surface tensions on either side of the barriers, since $\gamma_0 > \gamma$, when the area available to the monolayer is reduced the surface density of molecules increases, decreasing γ and increasing π. Thus the plot of π against area per molecule (A) increases as the available area is decreased, and is the two-dimensional analog of a pressure–volume isotherm (see Fig. 2.2). Surface pressure–area (π-A) isotherms reflect both the structure of the molecules in the monolayer and the interactions between them (see sec. 2.3). Two parameters that are commonly obtained from the π-A isotherm are the limiting area per molecule, obtained by extrapolation of the steep part of the isotherm to zero surface pressure, and the surface pressure at which the monolayer buckles out of the interface and the surface pressure drops, the so-called collapse pressure. For close-packed (solid) monolayers of normal fatty acids, the limiting areas on acid subphases are independent of the alkyl chain length. For example, stearic acid

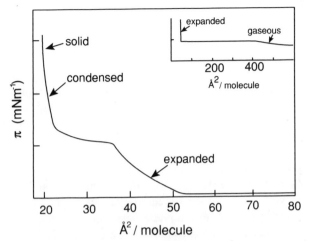

Fig. 2.2. Surface pressure (π)–area per molecule (A) isotherm for a typical fatty acid (e.g., pentadecanoic acid $C_{14}H_{29}CO_2H$) at the aqueous–air interface. The inset shows the region of coexistence of the gaseous and liquid expanded states. From Knobler (1990a). Copyright 1990 John Wiley & Sons.

($C_{17}H_{35}CO_2H$) and hexatriacontanoic acid ($C_{35}H_{71}CO_2H$) have the same limiting area, approximately 20 A^2 per molecule (0.2 nm^2 per molecule), which is a measure of the cross-sectional area of the alkyl chains and shows that they are arranged perpendicularly to the interface. Their collapse pressures are significantly different—42 mNm^{-1} for stearic and 58 mNm^{-1} for hexatriacontanoic acid—which reflects the fact that there is more cohesion between the close-packed longer alkyl chains and a greater surface pressure is required to buckle the monolayer.

Besides the Langmuir trough technique, several other physical methods have been used to investigate monolayers, including surface potential and surface viscosity techniques. In the surface potential technique, and air electrode—usually consisting of an insulated metal wire tipped with an α-emitter (e.g., americium-241) to ionize the air—is placed just above the monolayer, and a reference electrode (e.g., Ag/AgCl or calomel) is placed in the subphase. The potential between the electrodes is measured, using a high-impedance electrometer, in the presence (V_m) and absence (V_o) of the monolayer, and "surface potential" (ΔV) is defined by the equation

$$\Delta V = V_m - V_o \tag{2.2}$$

The surface potential reflects the polarity and orientation of surface dipoles, largely but not entirely due to the headgroups of amphiphiles in the monolayer. Thus for a surface dipole (μ) oriented at an angle θ to the perpendicular, ΔV

is given by

$$\Delta V = \frac{n\mu \; \cos\theta}{\epsilon_o \epsilon_r} \qquad (2.3)$$

where n is the number of molecules per unit area in the monolayer and ϵ_o and ϵ_r are the permittivity of vacuum and relative permittivity of the medium, respectively. There are considerable theoretical difficulties in the interpretation of surface potential measurements, since $\mu \; \cos\theta$ and ϵ_r are not usually precisely known.

In recent years a number of optical methods have been applied in monolayer studies (Knobler, 1990a); these and the information that can be obtained from them, together with other physical methods, are summarized in Table 2.1. Of particular interest in relation to the phase behavior in monolayers is the use of fluorescence microscopy (Knobler, 1990b), which provides direct information on the morphology of insoluble monolayers. In this technique a fluorescent probe is introduced into the monolayer at a relatively low concentration ($\sim 1\%$); typical probes are 7-nitrobenz-2-oxa-1,3-diazol-4-yl linked to dipalmitoylphos-

TABLE 2.1. Techniques Used in the Study of Insoluble Monolayers

Techniques	Parameter Measured	Information Obtained
Surface pressure	$\pi(\text{mNm}^{-1})$ vs. A	Limiting area, collapse pressure, state of the monolayer
Surface potential	$\Delta V(\text{mV})$ vs. A	Surface dipole, headgroup orientation
Surface viscosity	η_s	Viscosity changes
Ellipsometry	Polarization of reflected light	Refractive index and thickness of the monolayer
X-ray diffraction (high intensity X-rays from synchroton sources)	Reflectivity	Lattice spacing and structure in high-density monolayers
Neutron diffraction	Reflectivity	As for X-rays, but the contribution of water to the reflectivity can be eliminated
Second-harmonic generation	Polarization of the second harmonic of laser light–nonlinear susceptibility (χ^2)	The average value of the polar angle between the molecular axis and the normal to the interface
Fourier transform (External reflection) Infrared spectroscopy)	Reflectivity	Average orientation of molecular chains in the monolayer
Fluorescence microscopy	Fluorescence from probe molecules	Phase behavior and monolayer structure

phatidylethanolamine (NBD-DPPE) and NBD-hexadecylamine. The probe is excited with a laser or arc lamp, and images of the monolayer are detected with a high-sensitivity television camera and recorded on videotape. Differences in the intensity of fluorescence within the monolayer can reflect difference in density between phases or differences in solubility.

2.3. STATES OF MONOLAYERS

Surface pressure/area per molecule isotherms extending over a wide range of area per molecule for fatty acids or phospholipids show a number of distinct regions that correspond to different monolayer states and regions of coexistence of two states. Figure 2.2 shows a typical isotherm for a long-chain fatty acid. A number of states can be clearly identified as the monolayer is compressed (i.e., as the area per molecule at the interface is reduced); starting with the expanded gaseous state (G), G + the liquid-expanded state (LE), LE, LE + the liquid-condensed state (LC), and finally LC and the solid (S) state. The four distinct states G, LE, LC, and S correspond to decreasing headgroup separations and increasing ordering and loss of conformational freedom of the alkyl chains. In the region of coexistence of the G + LE states, the surface pressure is constant, consistent with a first-order phase transition from the G to the LE state. The nature of the transition from the LE to the LC state has been a contentious issue for many years. There are in the literature many isotherms showing the surface pressure rising in the region of coexistence of the LE and LC states, which have been presented as evidence for a degenerate or "higher order" phase transition between the two states. However, Pallas and Pethica (1985) have shown that for high-purity fatty acids or phospholipids, provided the compression rate is not too high and the relative humidity is controlled between 98 and 100%, the surface pressure remains constant through the region of coexistence of the LE and LC states, proving that the LE to LC transition is first-order. Evidence for the coexistence of two phases at constant surface pressure can be found from surface potential measurements. Using an electrode with a 1 cm diameter, fluctuations in the surface potential in the LE + LC regions are found, which suggests that the inhomogeneities in the monolayer density causing the fluctuations are at least of the order of millimeters.

The morphology of coexistent monolayer states can be revealed using fluorescence microscopy (Knobler, 1990a,b). The fluorescent probe NBD-hexadecylamine can be incorporated into monolayers at a concentration of about 1%. When excited, the probe density and/or degree of quenching in different monolayer states reveals the morphology of the monolayer. The fluorescence is high in the LE state and low in the G and LC states. Figure 2.3 shows the images produced by fluorescence microscopy on compressing a pentadecanoic acid monolayer from the two-phase G + LE region to the LE + LC region. The gas phase "bubbles" appear black in contrast to the LE state and shrink as the monolayer is compressed, until a uniform LE phase is formed. Further

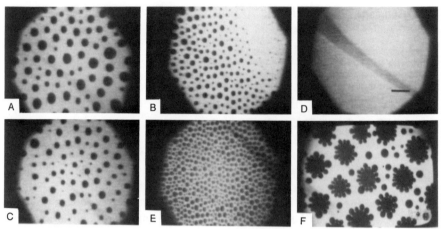

Fig. 2.3. Fluorescence microscope images of pentadecanoic acid monolayers at increasing extents of compression. **A.** gaseous phase "bubbles" in liquid-expanded monolayer (G + LE); average area per molecule $61A^2$. **B.** G + LE after compression to $51A^2$ per molecule. **C.** LE monolayer at $36A^2$ per molecule. **D.** liquid-condensed (LC) "black" regions in LE monolayer; average area per molecule $27A^2$. **E.** LE + LC after compression to $24A^2$ per molecule. **F.** image formed by cooling a monolayer in the two-phase region (G + LE) until circular LC regions form and the G bubbles accumulate around them. Note the bar in **C** corresponds to 100 μm. From Knobler (1990b), with permission of the publisher and the author.

compression results in the formation of areas of LC state (which appear black). These areas increase in number as the monolayer is compressed, until the entire monolayer is in the LC state and the field appears uniformly black.

The coexistence of three phases can be observed if a two-state monolayer, G + LE, is cooled at constant area per molecule. Figure 2.3F shows the fluorescent image seen when a G + LE monolayer of pentadecanoic acid is cooled below 17°C. Circular domains of the LC state are formed surrounded by "bubbles" of G state when the monolayer is cooled below the G-LE-LC triple point, as shown in the temperature/area per molecule phase diagram in Figure 2.4. The LE state is unstable below the triple point, but the rate of transformation is slow so that the coexistence of the three states can be observed. The LE + G region of coexistence ends in a critical point at a higher temp/ ature, but as yet there is no evidence of a critical point for the LC + LE region.

It should be noted that the use of fluorescent probes is of course subject to the same criticisms of all probe analysis, in that the probe is essentially an impurity in the system. While effects have been observed in monolayers that have been attributed to the presence of the probe (Knobler, 1990a)—specifically the formation of complex patterns resembling the fractal patterns that occur on supercooling—these are essentially nonequilibrium effects. Despite these short-

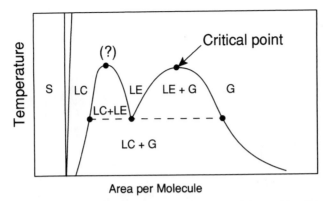

Fig. 2.4. Temperature/area per molecule phase diagram. The width of the LC + LE region has been exaggerated compared to the LE + G region. The existence of an LC + LE critical point has not yet been proven. Adapted from Knobler (1990b).

coming, the use of fluorescence imaging has greatly added to our visualization and knowledge of monolayer states and their behavior.

2.4. PHOSPHOLIPID MONOLAYERS

Phospholipid monolayers exhibit many of the features described above when either the acyl chain length or temperature is varied. Figure 2.5 shows the surface pressure–area isotherms for a homologous series of phosphatidylcholines (PCs or lecithins). The shorter acyl chain compounds, dicapryl and dimyristoyl, give LE state monolayers, while the long-chain compounds, dibehenoyl and distearoyl, give LC state monolayers at 22°C. Dipalmitoyl PC shows a transition from the LE to the LC monolayer. The curve is representative of many isotherms given in the literature, although from the work of Pallas and Pethica (1985) the isotherm should have a horizontal section joining the LE and LC regions. In practice, it is extremely difficult to achieve a sufficiently high degree of purity to give a constant surface pressure over the region of coexistence of the LE + LC states. It is seen that only the one homologue gives a two-state monolayer at 22°C, and this is the dipalmitoyl PC, which has a chain-melting temperature of approximately 41°C. The chain-melting temperature in the bulk phase of a phospolipid corresponds to the transition between the gel state (L_β') bilayer and the liquid-crystalline state (L_α). In the gel state, the acyl chains are in an all-*trans* conformation. On heating, *trans*-gauche conformational changes about the C—C bonds are introduced so that the acyl chains effectively melt, while the long-range order between the molecules in the bilayers is retained.

There is a correspondence between the behavior of phospholipids in the bulk

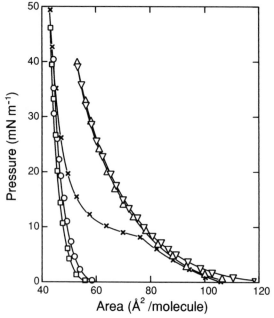

Fig. 2.5. Pressure-area curves for saturated lecithins on 0.1 M NaCl at 22°C. $\square$, dibehenoyl (C_{22}); $\bigcirc$ distearoyl (C_{18}); $\times$, dipalmitoyl (C_{16}); $\triangle$, dimyristoyl (C_{14}); ∇, dicapryloyl (C_6). From Phillips and Chapman (1968), with permission of the publisher.

phase and their behavior in monolayers. Dibehenoyl and distearoyl PCs have chain-melting temperatures of approximately 75°C and 54°C, respectively, and form LC monolayers at 22°C, while dimyristoyl and dicapryloyl PCs have chain-melting temperatures of 24°C and < 0°C, respectively, and form LE monolayers at 22°C.

For a particular phospholipid, the surface pressure–area isotherms show a change from the LC state to the LE state with increasing temperature (Fig. 2.6) (see also Chapter 4). The region of coexistence of the LC + LE phases decreases with increasing temperature. In the two-phase region, small but definite fluctuations in surface potential have been observed that were consistent with a first-order transition between the two phases (Pallas and Pethica, 1985). The changes in the monolayer structure on going from the LE to the LC states has been studied by specular neutron reflection techniques (Bayerl et al., 1990). Neutrons are specularly reflected in the same manner as light polarized perpendicular to the plane of reflection, so that the reflectivity of the surface can be considered in terms of a number of layers of defined thickness and refractive index. The refractive index (n) for neutrons is a simple function of the composition of the layers.

$$n = 1 - \left(\frac{\lambda^2}{2\pi}\right) \Sigma n_i b_i \qquad (2.4)$$

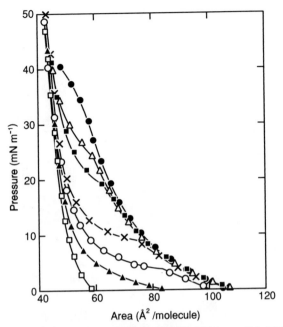

Fig. 2.6. Pressure-area curves for L-α-dipalmitoyllechithin on 0.1 M NaCl at various temperatures. ●, 34.6°, △, 29.5°; ■, 26.0°; ×, 21.1°; ○, 16.8°; ▲, 12.4°; □, 6.2°. From Phillips and Chapman (1968), with permission of the publisher.

where λ is the wavelength of the neutron beam (0.5 to 6.5 Å), n_i the number density, and b_i the scattering length of the nucleus i. (Lee et al., 1989). For neutrons, the scattering lengths of deuterium and hydrogen (and hence of D_2O and H_2O) are of opposite sign, and by taking an appropriate H_2O/D_2O mixture, molar ratio 0.088, with the same scattering length density as air, the specular reflection from the subphase can be eliminated (the so-called contrast matched water [CMW] condition). If pure D_2O is used as subphase a strong contrast to air is obtained. Using a single-layer model for the specular neutron reflection from deuterated dimyristoylphosphatidylcholine (DMPC) monolayers containing a small proportion of negatively changed phospholipids, the data shown in Table 2.2 were obtained for the monolayer in the LE and LC states. In the LC state, the measurements of the scattering-length density and thickness of the monolayer are the same for the two methods of measurement (CMW and D_2O subphase), but in the expanded state (LE) the D_2O subphase method gives a reduced thickness at low surface pressure. This implies that the phospholipid headgroup penetrates into the subphase. This penetration is not seen using the CMW method at low surface pressure, because of the zero contrast between the subphase and the air. The conformational change in the glycerol backbone region of the phospholipid on going from the LC state to the LE state is shown in Figure 2.7. The deeper penetration of the headgroup region of the molecule results in greater headgroup hydration, but as well as this the acyl chains in

TABLE 2.2. Results of One-Layer Fits of the Neutron Specular Reflection Data of DMPC/DMPG (7:3) Monolayers on a Contrast Matched Water (CMW) Subphase and on a D_2O Subphase

Phase State	Subphase	Lateral Pressure π (mN/m)	Scattering Length Density ($n_{b,fit}$, 10^6Å^{-2})	Monolayer Thickness (d_{fit}, Å)	Molecular Area ($A_{(\pi)}$, Å^2)	Molecular Volume ($V_m = d_{fit}A_\pi$, Å^3)
Expanded	CMW	11	2.38 ± 0.4	19.5 ± 1	66.0	1,280
Condensed	CMW	31	3.24 ± 0.4	22.5 ± 1	53.0	1,192
Expanded	D_2O	6	3.42 ± 0.4	15.5 ± 1	72.0	1,101
Condensed	D_2O	27	3.18 ± 0.4	22.8 ± 1	53.0	1,208

From Bayerl et al. (1990), with permission of the publisher.

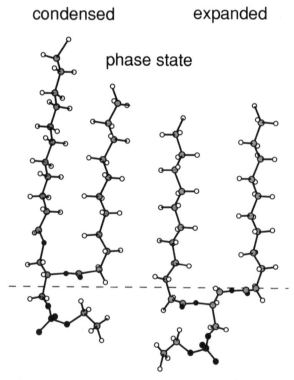

Fig. 2.7. Conformational changes in the glycerol backbone region of the phospholipids in a monolayer at the transition from the expanded (fluid) to the condensed (solid) phase state. The dotted line indicates the border up to which water from the subphase can penetrate into the headgroup region. In the condensed state, the acyl chains are more ordered and tilted at approximately 32° to the vertical (those details not shown) From Bayerl et al. (1990), with permission of the publisher.

the condensed state are more ordered and are tilted at an angle of approximately 32° to the vertical. The change in monolayer thickness from 22.5 Å to 19.5 Å (CMW method), on passing from the LC to the LE state (Table 2.1), compares with estimates of thickness of one monolayer leaflet in orientated multilayers as measured by neutron diffraction of 22.0 Å (L'_β-state) to 19.5 Å. Thus we have in this system a clear correspondence between the monolayer behavior and that in the lyotropic mesomorphic states.

2.5. PHOSPHOLIPID MONOLAYER TO SURFACE BILAYER TRANSITION

In 1979 Tajima and Gershfeld reported that they had observed the spontaneous formation of a bilayer film at the air–water interface. The transition from a condensed monolayer to a surface bilayer occurs at a singularity in temperature that is characteristic of the phospholipid. Figure 2.8 shows the surface prop-

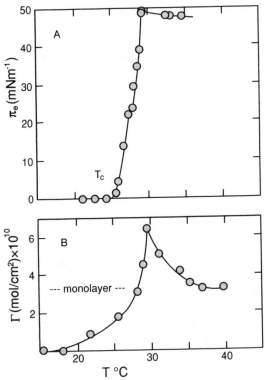

Fig. 2.8. The surface properties of DMPC dispersions in water are shown as a function of temperature. **A.** Equilibrium surface pressure, π_0. **B.** Surface concentration, Γ. For condensed monolayers, A = 55 Å²/molecule. From Tajima and Gershfeld (1985), with permission of the publisher.

erties of a dispersion of DMPC in water as a function of temperature. Both the surface concentration (Γ) and the equilibrium surface pressures (π_e) rise dramatically as the temperature is raised and go through maxima at 29°C, a few degrees above the chain-melting temperature of DMPC (23.5°C).

The surface concentration at 29°C is twice that found at 37°C, at which DMPC forms a condensed monolayer with an area per molecule of 55A^2, and hence it is deduced that at 29°C a surface bilayer is formed that is equivalent to two condensed monolayers (Tajima and Gershfeld, 1985). Surface bilayer formation has also been observed for dioleoyl PC, DPPC, DMPG, and mixed dispersions of DMPC + dioleoyl PC. In each case, the surface pressure maximum for the pure phospholipids occurs a few degrees above the gel to liquid-crystalline transition temperature for the bulk dispersion. In the case of the mixed phospholipid dispersion, the surface pressure and concentration have maxima between the transition temperatures of the two components.

The transition from the monolayer to the surface bilayer is not a first-order process, as there are no discontinuities in the π_e-T or Γ-T curves. The Γ-T curves change continuously over the temperature interval where the surface bilayer phenomenon is seen. In the case of mixed phospholipid dispersions, the surface bilayer has the same composition as the bulk state lipid mixture. The structure of the surface bilayer is equivalent to two condensed monolayers which, in the case of DMPC, have an area per molecule of 55A^2, which is close to the area per molecule in multilamellar dispersions of lipid.

2.6. ION INTERACTIONS AND ELECTROSTATIC EFFECTS IN PHOSPHOLIPID MONOLAYERS

The composition of the aqueous subphase can have significant effects on the behavior of monolayers. The pH can affect the degree of ionization of phospholipid headgroups and hence the electrostatic interactions between them. Surface potential measurements have shown that for the common phospholipids PC, PE, PA, PS, and PI, the surface potential increases as the phosphate residue is protonated in the pH range 4 to 1. At alkaline pH, the surface potentials of PE and PS decrease as the $-NH_3^+$ groups are deprotonated. PA behaves similarly as the second acidic proton is removed (Papahadjopoulos, 1968) on an alkaline subphase. In the pH range of approximately 6–8, the states of ionization of these common phospholipids remain unchanged and the surface potentials are constant. As a general rule, increasing the headgroup charge will lead to greater lateral electrostatic interactions and move expanded monolayers. Increasing ionic strength by addition of salts with monovalent counterions (i.e., the ions of opposite charge to the headgroup) will shield the headgroups' charges and lead to more condensed monolayers (Helm et al., 1986).

The effects of multivalent ions on phospholipid monolayers are more complex than those of monovalent ions. Of particular interest are the divalent cations Ca^{2+} and Mg^{2+}, which have dramatic effects on the monolayers of phospholipids with negatively charged headgroups. Figure 2.9a shows the effect of Ca^{2+} concentrations in the subphase on the surface pressure–area isotherms of dimyristoylphosphatidic acid (DMPA). In the absence of Ca^{2+} ions, this phospholipid shows the typical transitions from the G to LE and LE to LC states. Addition of increasing Ca^{2+} ion concentration in the subphase at micromolar levels induces a condensation of the monolayer, and both the surface pressure and area at which the LE to LC transition occurs are substantially reduced; at a Ca^{2+} concentration of 50 μM the transition is eliminated. Mg^{2+} ions behave qualitatively similarly, although the reduction in surface pressure is not as large and the LE to LC phase transition is still observed at a pMg of 3. Figure 2.9b shows the effects of Ca^{2+} and Mg^{2+} ion concentration on the surface pressure at the LE–LC transition point (π_c) and the area per molecule at the G–LE transition point (A_{fl}). Both Ca^{2+} and Mg^{2+} ions reduce the area per molecule at which the G–LE transition occurs due to ion binding and reduce the surface pressure at which the LE monolayer undergoes the LE–LC transition.

The morphology of DMPA monolayers in the presence of divalent ions can be revealed by fluorescence imaging. On increasing the Ca^{2+} ion concentration in the subphase between 1 and 50 μM, the LC domains, which contain the fluorescent probe NBD-DPPE and appear as dark areas against the more fluorescent LE phase, change from being approximately circular to being irregular and eventually dendritic, as shown in Figure 2.10. In the presence of only monovalent ions, the LC domains are circular and form (approximately) a hexagonal superlattice due to electrostatic repulsion; the addition of Ca^{2+} ions, by binding to the phosphatidic acid headgroups, reduces the repulsion and allows more chaotic (dendritic) surface structures to form.

The influence of electrostatic interactions on the uniformity of size of the domains in a monolayer can be considered in terms of the line tension (γ) and the density of surface dipoles (ρ_D). At a given mean area per molecule and fraction of LC phase, the Gibbs energy change due to a fluctuation in an area f_o results in one domain increasing by Δf at the expense of another domain decreasing by $-\Delta f$. Thus

$$\Delta G = [G(f_o + \Delta f) - G(f_o)] + [G(f_o - \Delta f) - G(f_o)] \qquad (2.5)$$

$$= \left(\frac{\partial^2 G}{\partial f^2}\right)_{f_o} (\Delta f)^2 \qquad (2.6)$$

It has been shown (Lösche and Möhwald, 1989) that

$$\left(\frac{\partial^2 G}{\partial f^2}\right) = -\frac{\gamma}{f_o^{3/2}} + \left(\frac{\rho_D^2}{4\pi\epsilon_0\epsilon_r} \times \frac{\sqrt{5}}{2f_o^{3/2}}\right) \qquad (2.7)$$

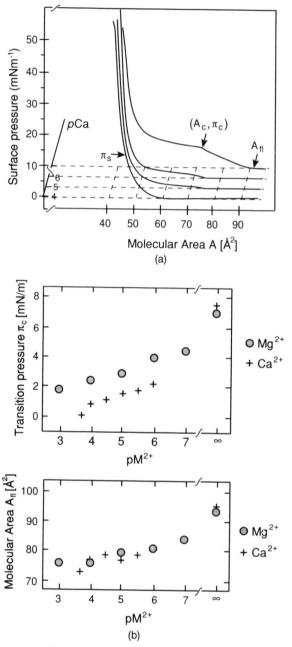

Fig. 2.9. a. Effect of Ca^{2+} ion concentration on DMPA monolayers at constant pH (= 6) and temperature (= 27.5°C). A_{fl} denotes the G to LE transition, A_c (and π_c) denote the LE to LC transition, and π_s denotes the LC to S transition. **b.** Effect of Ca^{2+} and Mg^{2+} concentrations on π_c and A_{fl} at pH 6 and 27.4°C (Mg^{2+}) and 28.1°C (Ca^{2+}). From Lösche and Möhwald (1989), with permission of the publisher.

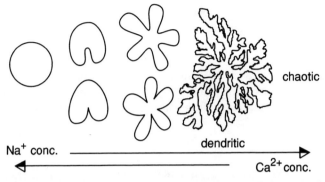

Fig. 2.10. Characteristics of domain shapes observed with DMPA monolayers at pH 6. The arrows indicate directions into which the shapes are transformed on increasing Ca^{2+} or Na^+ concentrations. From Lösche and Möhwald (1989), with permission of the publisher.

The sign of the second (electrostatic) term in equation 2.7 is opposite to that of the line tension term and generally exceeds it. Hence ΔG will be positive, favoring a uniform domain size. In the presence of divalent ions, the electrostatic forces are reduced due to a reduction in ρ_D, and domains can then grow irregularly.

2.7. MIXED MONOLAYERS

The fact that biological membranes are never composed of a single lipid component has stimulated many studies on mixed monolayers in an attempt to understand the nature of the interactions between different membrane lipid components. The range of lipid mixtures that can now be studied is virtually infinite, and as yet even binary combinations of lipids have not been completely investigated. In thermodynamics terms, the treatment of binary monolayer systems is the two-dimensional analog of the treatment of mixing two components in the bulk phase. In the absence of specific interactions when the two components form an ideal (random) mixture, the average area per molecule in the mixture ρ_{12} is simple the sum of the areas occupied by each of the components, so that at a given surface pressure

$$\rho_{12}(\text{ideal}) = x_1\sigma_1 + x_2\sigma_2 \tag{2.8}$$

where x_1 and x_2 are the mole fractions of components 1 and 2 with areas per molecule of σ_1 and σ_2, respectively. Thus since $x_1 + x_2 = 1$, σ_{12} will be linear in x_1

$$\sigma_{12}(\text{ideal}) = x_1\sigma_1 + (1 - x_1)\sigma_2 = x_1(\sigma_1 - \sigma_2) + \sigma_2 \tag{2.9}$$

Any deviation from this ideal behavior will result in loss of linearity between σ_{12} and x_1. Repulsive interactions will give rise to positive deviations ($\sigma_{12} > \sigma_{12}$[ideal]) and attractive interactions to negative deviations ($\sigma_{12} < \sigma_{12}$[ideal]).

Deviations from ideality can be expressed in term of excess Gibbs energies of mixing, as in the treatment of mixing in the bulk state (Goodrich, 1957). A small change in Gibbs energy at constant composition, temperature, and pressure arising from a small change in surface tension is given by

$$dG = -\sigma d\gamma \qquad (2.10)$$

If we consider two "unmixed" monolayers being compressed from a surface tension γ^* almost equal to that of the subphase (γ_o) to a surface tension γ, the change in Gibbs energy per mole of lipid is given by

$$G - G^* = -X_1 \int_{\gamma^*}^{\gamma} \sigma_1 d\gamma - X_2 \int_{\gamma^*}^{\gamma} \sigma_2 d\gamma \qquad (2.11)$$

If the monolayers are allowed to mix at surface tension γ^*, it may be assumed that as the area per molecule at this surface tension is very large, mixing is ideal; hence

$$G_{12}^* - G^* = RT\, (X_1 \ln X_1 + X_2 \ln X_2) \qquad (2.12)$$

If such a mixed monolayer is compressed from surface tension γ^* to γ, then

$$G_{12} - G_{12}^* = -\int_{\gamma^*}^{\gamma} \sigma_{12} d\gamma \qquad (2.13)$$

Addition of equation 2.12 and 2.13 and subtraction of 2.11 gives

$$G_{12} - G = -\int_{\gamma^*}^{\gamma} \sigma_{12}\, d\gamma + X_1 \int_{\gamma^*}^{\gamma} \sigma_1\, d\gamma + X_2 \int_{\gamma^*}^{\gamma} \sigma_2\, d\gamma$$
$$+ RT\, (X_1 \ln X_1 + X_2 \ln X_2) \qquad (2.14)$$

Hence the excess Gibbs energy of mixing is given by the first three terms on the right-hand side of equation 2.14 which on changing the limits of integration may be written

$$\Delta G_m^{xs} = \int_{\gamma}^{\gamma^*} (\sigma_{12} - X_1\sigma_1 - X_2\sigma_2)\, d\gamma \qquad (2.15)$$

In terms of surface pressure, $\pi = \gamma_o - \gamma$ and $\pi^* = \gamma_o - \gamma^*$ and $d\pi = -d\gamma$.

Hence

$$\Delta G_m^{xs} = \int_{\pi^*}^{\pi} (\sigma_{12} - X_1\sigma_1 - X_2\sigma_2) \, d\pi \qquad (2.16)$$

The excess Gibbs energy of mixing can thus be calculated by integration under the π–σ isotherms of the mixed monolayer and the pure components from zero surface pressure to π, since when $\gamma^* \approx \gamma_o$, $\pi = 0$.

Ideal behavior occurs in mixed monolayers of phospholipids only when the acyl chain lengths of the two components differ by less than two carbon atoms; e.g., DSPC + DPPC exhibit ideal behavior at 22°C regardless of whether the surface pressure is such that pure DPPC would be in the LE or LC states. Mixtures of phospholipids differing in degree of saturation—e.g., DSPC + dioleoyl PC, where the acyl chain lengths are the same (C_{18})—exhibit positive deviations from ideal behavior at 22°C. If the acyl chain lengths differ greatly in chain length—e.g., dibehenoyl (C_{22}) PC + dioleoyl (C_{18}) PC—the phospholipids become immiscible, and under these circumstances the average area per molecule may be given by equation 2.8 and the mixed monolayer appears ideal, although in reality it is heterogeneous.

Figure 2.11 shows the surface pressure–area isotherms for mixed monolayers of DPPC + phosphatidylinositol (PI) of several compositions. Addition of PI

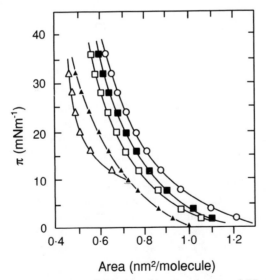

Area (nm²/molecule)

Fig. 2.11. Surface pressure–area isotherms for monolayers of PI, DPPC, and their mixtures at the aqueous–air interface at 25°C. The substrate was 20 mM imidazole, pH 6.6, ionic strength 0.031 M. ○, PI; ■, PI-DPPC (weight fractions, 0.75:0.25); □, PI-DPPC (0.5:0.5); ▲, PI-DPPC (0.25:0.75); △, DPPC. From Cordoba et al. (1990), with permission of the publisher.

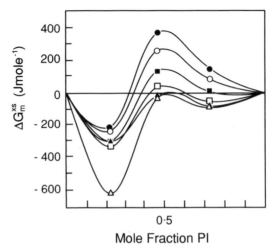

Fig. 2.12. Excess Gibbs energies of mixing in PI-DPPC monolayers on imidazole buffer (20 mM, pH 6.6, ionic strength 0.013 M). The surface pressures were: $\triangle$, 5 mN m^{-1}; $\blacktriangle$, 10 mN m^{-1}; $\square$, 15 mN m^{-1}; $\blacksquare$, 20 mN m^{-1}; $\bigcirc$, 25 mN m^{-1}; $\bullet$, 30 mN m^{-1}. From Cordoba et al. (1990), with permission of the publisher.

(from wheat germ) to DPPC results in the loss of the LE to LC transition point. Treatment of this data by the method described above gives the Gibbs energies of mixing and excess Gibbs energies as a function of monolayer composition, shown in Figure 2.12. Here the deviations from ideality are negative at low mole fraction PI and positive at high mole fraction. At low surface pressures at a mole fraction of 0.5, the monolayer behaves almost ideally.

Mixed monolayers of DPPC and phosphatidylglycerols having unsaturated acyl chains, such as egg phosphatidylglycerol, show more complex surface pressure–area isotherms that have a direct relevance to the behavior of pulmonary surfactants found at the air–aqueous interface in the alveoli of the mammalian lung. The pulmonary surfactant is a protein-lipid complex rich in phospholipids (80–85%) of which DPPC is a major component and PG is present at a relatively high concentration. The function of pulmonary surfactant is to reduce the surface tension at the air–aqueous interface in the lung and prevent the collapse of the alveoli at the end of expiration, and to decrease the pulmonary effort during breathing. A deficiency or alteration in lung surfactant, which can occur in premature babies, results in respiratory distress syndrome (RDS). RDS studies have shown that it is associated with decreased levels of PG in the surfactant of premature babies. The surface properties of pulmonary surfactant are unusual in that at body temperature the surfactant spreads readily at the air–water interface to form a monolayer that can be compressed to a high surface pressure of over 60 mNm^{-1}. However, although DPPC has a high collapse pressure ($\sim$70 mNm^{-1}), below its chain-melting temperature ($\sim$41°C) it will only spread above T_c and then the collapse pressure is significantly lower

(~ 50 mN m^{-1}). To explain how pulmonary surfactant can spread readily below the T_c of DPPC while having a high collapse pressure characteristic of the L_c state, the concept of "preferential squeeze-out" was introduced by Morley and Bangham (1981), following on from the concept of "self-purification" by Clements (1977). The principal idea is that the presence of components such as PG in the surfactant lowers the T_c below body temperature so that the surfactant spreads, but on compression the T_c-lowering components are squeezed out of the monolayer, which then becomes rich in the phosphatidylcholines and can then exhibit a high collapse pressure.

Studies on mixtures of DPPC and egg PG have demonstrated that such behavior can be observed, provided that the PG contains unsaturated acyl chains (de Fontanges et al., 1984; Boonman et al., 1987). Figure 2.13 shows a schematic plot of surface pressure–area isotherms of DPPG–egg PG monolayers that demonstrates their behavior on compression. When first compressed, the monolayer shows the characteristic LE to LC transition, but on further compression a plateau region is reached at ~ 48 mN m^{-1}, followed by a further rise in surface pressure before the monolayer collapses at 70 mN m^{-1}. In the plateau region, lipid molecules are lost from the monolayer, PG being preferentially but not exclusively squeezed out. The loss of molecules from the monolayer increases with the concentration of DPPC and decreases with the rate of compression. The occurrence of the plateau depends on the loss of monolayer

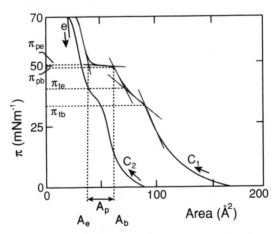

Fig. 2.13. Schematic plot of surface pressure (π) vs, surface area (A) of DPPC-egg PG monolayers. Abbreviations and symbols: t, LE–LC transition; A_b and A_e, surface areas at the onset and at the end, respectively, of the plateau formation; A_p, surface area swept by the barrier during plateau formation; π_{tb} and π_{te}, surface pressure at the onset and end, respectively, of the LE–LC transition region; π_{pb} and π_{pe}, surface pressure at the onset and end, respectively, of the plateau formation; C_1 and C_2, first and second compressions, respectively; e, expansion. The onset and the end of the transition regions are found by constructing the intersections of tangents of the curve. From Boonman et al. (1987), with permission of the publisher.

in the LE state from the interface, i.e., it occurs before all the LE monolayer in the LE-LC two-phase region has been converted to the LC state. After compression to the collapse point, if the monolayer is reexpanded and a second compression carried out, the surface pressure–area curve is shifted to lower areas because fewer molecules are present at the interface and a plateau is no longer present due to the preferential loss of PG from the monolayer.

Mixtures of phospholipids and cholesterol have always been of great interest due to the occurrence of cholesterol in cell membranes, often at high concentration, as for example in the human erythrocyte membrane (~ 42 mole %). It is well established that cholesterol generally has a condensing effect in mixed monolayers with phospholipids, so that the average area per molecule in the mixed monolayer is less than would be calculated from equation 2.9 for an ideal mixture of the two components. The possible formation of molecular complexes, with phospholipid to cholesterol molar ratios of $2:1$, $1:1$, and $1:2$, which could account for the condensing effect of cholesterol, have received much attention. In the bulk phase, it is found that cholesterol broadens and eventually eliminates the gel to liquid-crystalline phase transition of phospholipids as the mole % is increased up to around 30. It is for this reason that the phase transition cannot in general be observed in plasma membranes of mammalian cells when the cholesterol content exceeds around 30 mole %, in contrast to bacterial cell membranes that contain no cholesterol, where phase transitions can be both observed and manipulated by controlling the fatty acid composition of the growth media and hence the acyl chain composition of the membrane lipids. Studies of the effect of cholesterol on the high-resolution NMR spectra of dispersions of phospholipid liposomes in D_2O show that cholesterol has a marked effect on the line width of the methylene protons of the phospholipid acyl chains. Spectral line widths reflect the rigidity or fluidity of the environment, narrow lines being associated with a fluid environment and rapid molecular motion, broad lines being associated with rigidity and slower molecular motion. Cholesterol is found to cause a narrowing of the line widths from methylene protons in phospholipid dispersions below their chain-melting temperature, but a broadening of line widths above their chain-melting temperature. Its effect is thus to fluidize acyl chains in the gel state and rigidify acyl chains in the liquid-crystalline state, so that the rigidity/fluidity of the acyl chains is largely independent of temperature at sufficiently high cholesterol levels. Cholesterol thus acts as a "plasticizer," keeping the bilayer fluidity uniform with changing temperature.

Cholesterol forms a solid monolayer when spread at the aqueous interface; the surface pressure–area isotherms are independent of temperature in the range 15–45°C and the limiting area is 0.39 nm^2 per molecule. The monolayer is 1.6 ± 0.5 nm thick, consistent with a near-vertical orientation of the cholesterol molecule at the interface (Baglioni et al., 1985). The development of fluorescence microscopy has enabled the morphology of mixed monolayers of cholesterol and phospholipids to be studied, revealing some surprising phenomena. For example, when mixed with DMPA in the doubly charged state on an alkaline substrate (pH 11), cholesterol has little effect on the LE-LC transition up to 5 mole %, slightly broadens it between 5 and 15 mole %, but beyond 15 mole % increases the broadening rapidly until at 25 mole % the transition is no longer detectable. The surface pressure at the LE-LC transition point is continuously shifted to higher values with increasing cholesterol content, at a rate of about 0.25 mN m^{-1} per mole % cholesterol (Heckl et al., 1988). Using the fluorescent probe dipalmitoylnitrobenzoxadiazolphosphatidylethanolamine (NBD-DPPE), the surface texture of the mixed monolayer has been observed as a function of cholesterol content up to 20 mole % at comparable degrees of condensed phase formation (crystallization) (Fig. 2.14). The probe is soluble in the fluid phase but not in the solid (condensed) domains, so that the condensed phase species appear black against a lighter background of fluid phase. As the cholesterol content is increased, the condensed-phase structure goes from near-circular aggregates to more elongated aggregates, to spirals, and then circular open aggregates, as summarized in Table 2.3.

The interpretation of these observations is that below 5 mole % cholesterol, the cholesterol reduces the line tension between the fluid mixed phase and the nearly pure DMPA condensed phase. This view is consistent with the observation that the width of the spirals formed changes on compression of the monolayers and the degree of crystallization, at a fixed mole % cholesterol in the range from about 1 to 5 mole %. At higher cholesterol concentrations, the situation is complicated by the probable incorporation of cholesterol into the condensed phase, which eventually leads to its elimination and possible artifactual effects at high compression due to the high concentration of the fluorescent probe in the residual fluid phase. Nevertheless, the observations give a fascinating insight into the unusual texture of cholesterol-containing monolayers.

2.8. ADSORPTION OF POLYPEPTIDES AND PROTEINS AT THE AQUEOUS–AIR INTERFACE

2.8.1 Polypeptide Monolayers

Before consideration of the adsorption of proteins at the aqueous–air interface, it is appropriate to consider the adsorption of polypeptides. Monolayers of synthetic homo-polypeptides have been studied with the objective of trying to establish and understand some of the fundamental properties of these polymers at interfaces, with the hope that the information will constitute a background to the behavior of the more difficult problem of protein adsorption.

Polypeptides can be spread from either pure organic solvents or organic solvent mixtures at the liquid–air interface to form stable monolayers that, when compressed, form a close-packed array of molecules with properties fairly characteristic of a solid monolayer. Figure 2.15 shows typical surface pressure and surface potential versus area curves for poly L-leucine monolayers spread

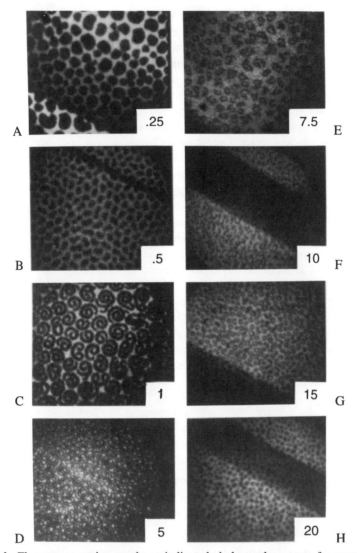

Fig. 2.14. Fluorescence micrographs at indicated cholesterol contents for comparable degrees of crystallization ϕ. **A.** 0.25%. **B.** 0.5%. **C.** 1%. **D.** 5%. **E.** 7.5%. **F.** 10%. **G.** 15%. **H.** 20%. From Heckl et al. (1988), with permission of the American Chemical Society.

from trifluoroacetic acid. For such a nonpolar polymer the force-area curves are independent of pH and ionic strength. The limiting area per amino acid residue lies in the range 0.166–0.176 nm^2 for this polypeptide, which compares with a value of 0.180 nm^2 residue^{-1} that is obtained by using X-ray diffraction on collapsed monolayer films that are known to be in the α-helical confor-

TABLE 2.3. Condensed-Phase Doubly Ionized DMPA/Cholesterol Structures at Different Cholesterol Concentrations

Mole % Cholesterol	Condensed-Phase Structure
0	Circular plus dendritic
0.25	Near-circular aggregates
0.5	Elongated aggregates
1.0	Spirals
7.5	Spirals (with frequent branching)
10	Spirals plus open circular
15	Circular open aggregates
20	Circular aggregates
25	Undetectable

From Heckl et al. (1988), with permission of the American Chemical Society.

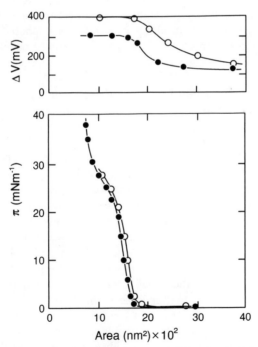

Fig. 2.15. Surface pressures (π) and surface potentials of poly-L-leucine monolayers at the air–water interface. $\bigcirc$, pH 2, I 0.01; $\bullet$, pH 7, I 0.02. From Phillips et al. (1979), with permission of the publisher.

mation. A close correlation has been found between the limiting areas per residue and those calculated from electron or X-ray diffraction for a range of other neutral polypeptides (Malcolm, 1973). Another feature of the pressure–area curves is the inflection at higher pressures, which is attributed to the formation of bilayers of α-helices formed on compression. The change in the surface potentials is due to changes in ionization of the terminal amino groups on compression.

By removing monolayers from the interface onto solid supports, it is possible to investigate them by spectroscopic methods, electron and X-ray diffraction, and hydrogen-deuterium exchange. The bulk of the available data is consistent with an α-helical conformation in the interface, at least for high molecular weight materials. In contrast, short-chain oligopeptides containing less than eight peptide residues probably have random conformations. This has been demonstrated using the fluorescent probe 2-p-toluidinylnaphthalene-6-sulphonate (TNS) (Tenenbaum et al., 1976). This probe is only fluorescent in a hydrophobic environment; when the radioactivity and fluorescence of tritiated TNS is monitored in monolayers of oligomeric γ-methyl-L-glutamates containing from 2 to 12 peptide units, transitions are found for chain lengths between 6 and 8 such that the radioactivity decreases and the fluorescence increases. Thus, although there are fewer TNS molecules in the interface for chain lengths above 8, their fluorescence is greater. The implication here is that at this chain length stable α-helices are formed, resulting in a more hydrophobic environment at the interface due to the presence of the γ-methyl groups on the surface of the helices. Below a chain length of 6, the oligomers have no well-defined secondary structure.

The α-helical conformation for polypeptides at the aqueous–air interface is not exclusive; for example, circular dichroism studies and multiple internal reflection measurements on Langmuir-Blodgett monolayers of poly γ-methyl-L-glutamate show that both α-helices and β-sheet conformations occur depending on the spreading solvent. Monolayers with the polypeptide largely in the β-conformation are found when the spreading solvent is pyridine + dichloracetic acid or pyridine + chloroform. In a β-sheet, the side chains are at right angles to the molecular axes with alternate side chains above and below the plane of the sheet. Two β-sheet structures are possible depending on whether the adjacent molecules are parallel (all the N termini at the same end) or antiparallel (alternate N and C termini at the same end). In a close-packed monolayer it is unlikely that we will get the exclusive formation of just one of these arrangements. As for the α-helix, there is considerable opportunity for both hydrogen bonding and hydrophobic interaction between the side chains. Whether a given polypeptide forms an α-helix or a β-sheet at the interface largely depends on the extent of hydrophobic interaction. For an α-helix having 18 residues in five turns, only about 4 residues are not involved in hydrophobic interactions, whereas in the β-conformation 9 will not be involved. This simple approach suggests that the α-helical conformation is favored; however, the fact that it is possible to produce both conformations with the same molecule clearly demonstrates that the energy difference between the two conformations at the interface is very small. The stereochemistry of the polypeptide also influences the conformation, as shown by a study of poly (L, D, and DL) alanine (Gabrielli et al., 1981). Multiple internal reflection measurements on monolayers transferred onto germanium plates show that poly (L and D) alanine are present at the water–air interface in the α-helix conformation, whereas poly (DL) alanine monolayers contain a considerable proportion of molecules in the β-conformation.

Polypeptides with ionizable side chains are water-soluble when the ionizable groups are fully charged but will form monolayers at certain pHs. Figure 2.16 shows the surface pressure and surface potential curves for poly-L-glutamic acid at different pHs and different ionic strengths. The curves illustrate several aspects of adsorption. First, at pH 5.4, at which the carboxylic side chain is fully ionized, the surface pressures are very low. Decreasing the pH to 2, at which the side chain is largely unionized results in pressure-area curves markedly dependent on ionic strength. At low ionic strength (I = 0.01), the limiting area per molecule is ~0.15 nm^2/residue, but on increasing the ionic strength to 0.1 a more expanded monolayer is formed with a limiting area ~0.22 nm^2/residue. Both these curves show inflections at higher pressures. There are two possible explanations for this behavior. If the polypeptide forms loops and tails that penetrate the subphase, increasing the ionic strength may cause these to be "salted-out," so giving a more expanded monolayer. Alternatively, the inflections in the curves might also be interpreted in terms of an increase in loop formation on compression.

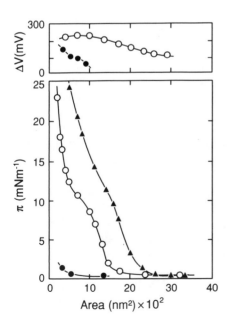

Fig. 2.16. Surface pressures (π) and surface potentials (ΔV) of poly-L-glutantic acid monolayers at the air–water interface. ○, pH 2, I 0.01; ▲, pH 2, I 2.0; ●, pH 5.4, I 0.02. From Phillips et al. (1979), with permission of the publisher.

2.8.2. Protein Monolayers

The surface activity of proteins was appreciated long before X-ray diffraction studies revealed their detailed structures, and the study of protein monolayers has attracted considerable attention over several decades and continues to be an active area of research. Much of the early work on protein monolayers was not easily interpreted in the absence of a detailed knowledge of protein structure, and it is only within the last two decades that the behavior of well-characterized proteins at interfaces has begun to be understood at a fundamental level.

The solubility of many proteins in aqueous media means that measurements can be made on the surface tensions of solutions using conventional techniques or, alternatively, proteins can be spread on a Langmuir trough and their surface pressure–area isotherms investigated. Because of the existence of secondary and tertiary structure in proteins, the question of the conformation at the interface and the extent of what is called surface denaturation are of central importance in the field. We can identify several problems associated with interfacial behavior of proteins.

1. The factors affecting the rate of adsorption from bulk solution to the interface.
2. The kinetics of conformational changes at the interface and their influence on subsequent adsorption, e.g., interpenetration by protein molecules in the subphase.
3. The factors controlling the extent of adsorption at equilibrium, e.g., the type of adsorption isotherm that is obeyed by a protein.
4. The conformation of the protein molecules in the equilibrium monolayer and the structure of the monolayer.

2.8.2.1. Kinetics of adsorption. The kinetics of adsorption of proteins at the air–water interface can be followed by several methods, including the measurement of surface tension as a function of time (t) or surface concentration from the change in surface radioactivity. The simultaneous measurement of surface pressure (π) and surface concentration (Γ) is a means of obtaining information about molecular rearrangements and penetration in the monolayer, factors that can be deduced from the correlation between π vs. t and Γ vs. t plots.

Figure 2.17 a and b show the π vs. t and Γ vs. t plots for β-casein (mol. wt. 24,000), a protein that has no rigid secondary or tertiary structure, and lysozyme (mol. wt. 14,306), a typical globular protein with a known tertiary structure. Both proteins were radiolabeled by acetylation of up to two of their lysine residues with [1-^{14}C] acetic anhydride. The proteins were injected into the subphase to an initial concentration $\sim 7 \times 10^{-5}$ wt.%. It is clear from these data that equilibrium surface properties are not reached for many hours at these concentrations.

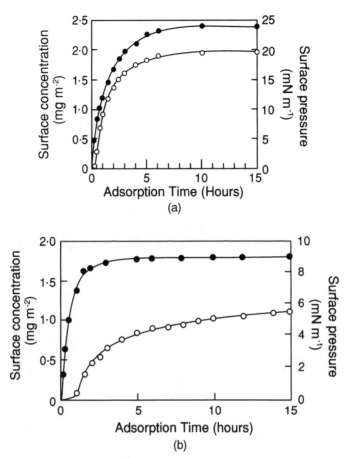

Fig. 2.17. Variation of surface pressure ($\bigcirc$) and surface concentration ($\bullet$) with time for the adsorption of β-casein (**a**) and lysozyme (**b**) at the air–water interface. Substrate phosphate buffer pH 7.1, I 0.01; protein concentration 7.3×10^{-5} wt % (**a**) and 7.6×10^{-5} wt % (**b**). From Graham and Phillips (1979), with permission of the publisher.

The kinetics of adsorption of the two proteins differ in that for β-casein π and Γ change simultaneously, whereas for lysozyme Γ reaches a steady value relatively quickly while the change in π exhibits an induction period and is still changing after 15 hrs. At high concentrations ($> 10^{-3}$ wt.%), the initial changes in π and Γ are too rapid to measure. The rate of change of π reflects the structural features of the protein; the more flexible the structure, the more rapidly the surface pressure comes to a constant value. A slow change in π is generally attributed to penetration of molecules into the monolayer and their subsequent rearrangement.

The kinetic features of adsorption can be categorized on the basis of the underlying mechanisms.

1. Diffusion-controlled adsorption. This will prevail during the initial stages of adsorption and will probably be irreversible. Both π and Γ will change with time (t) and the process should obey the equation.

$$\Gamma = 2C_p \left(\frac{Dt}{\pi}\right)^{1/2} \tag{2.17}$$

where C_p is the protein concentration in the subphase and D is the diffusion coefficient.

2. Adsorption with an energy barrier. This will be the situation after the surface pressure in the monolayer has exceeded some minimum value (~ 0.1 mN m^{-1}) and the adsorbing molecules have to do mechanical and/or electrical work in compressing the molecules already at the interface. Both π and Γ will change with time. The rate of change of the number of molecules (n) at the surface is given by the equation (MacRitchie and Alexander, 1963)

$$\frac{dn}{dt} = kC_p \exp\left[-(\pi\Delta A + q\Psi)/kt\right] \tag{2.18}$$

where k is a first-order rate constant, ΔA is the area required to be created against a constant surface pressure, π, to receive the molecule of charge q at the interface, which has an electrical potential Ψ in the plane of the charged groups. Equation 2.18 is based on the equations derived by Ward and Tordai (1946) with the assumption that the rate of desorption is zero (i.e., irreversible adsorption).

Diffusion-controlled adsorption has been demonstrated for numerous globular proteins, e.g., bovine serum albumin, β-lactoglobulin, ovalbumin, human γ-globulin, and myosin. For lysozyme and β-casein, plots of Γ vs. $t^{1/2}$ based on equation 2.17 are linear up to 1 hr when $C_p \sim 10^{-4}$ wt.%; however, the diffusion coefficients calculated from the slopes are an order of magnitude larger than those determined by dilute solution methods such as ultracentrifugation (this discrepancy has been attributed to convective stirring).

In the absence of an electrical barrier to adsorption, it follows from equation 2.18 that

$$\ln\left(\frac{dn}{dt}\right) = \ln(kC_p) - \pi\Delta A \tag{2.19}$$

hence from plots of $\ln(dn/dt)$ vs. π values of ΔA can be calculated. For the globular proteins mentioned above, it is found that ΔA falls in the range 1–1.75 nm^2 and is independent of the molecular weight of the protein. These figures are very much smaller than the cross-sectional area of the proteins and suggest that only a small part of the protein need penetrate the monolayer for adsorption to occur spontaneously. Ter-Minassian–Saraga (1981) has given an alternative interpretation and argues that ΔA is in fact related to the water activity at the interface and not the size of the absorbing molecules. Thus ΔA is proportional to the number of water molecules per protein molecule participating in protein denaturation and not the ''hole'' area created by the protein.

MacRitchie and Alexander (1963) investigated the predictions of equation 2.18 with regard to the electrical barrier to adsorption by measurement of the rate of adsorption (dn/dt) of lysozyme into monolayers of different electrical potential as reflected by their ξ-potentials. Thus the rate of adsorption of lysozyme (net charge $+ 9 \pm 1$ at pH 6.5) into a polyglutamic acid monolayer ($\xi = -51.5$ mV) is 44 times larger than into a polylysine monolayer ($\zeta = +38.5$ mV).

As Figure 2.17b illustrates, significant changes in π can occur when the surface coverage has reached saturation. Such changes reflect molecular rearrangement at the interface that can be analyzed in terms of the first-order equation.

$$\ln \left(\frac{\pi_{ss} - \pi_t}{\pi_{ss} - \pi_o} \right) = -t/\tau \tag{2.20}$$

where π_o, π_t and π_{ss} are the surface pressures at time t = o, t, and under steady-state conditions, respectively, and τ is a relaxation time. Analysis of the adsorption data for several globular proteins according to equation 2.19 shows the existence of two relaxation times, only the second relaxation process that occurs above t $\sim$ 4 hr gives the relaxation time (τ_2) for molecular rearrangements at constant Γ. The values of τ_2 depend on the protein concentration in the subphase and for lysozyme, β-casein, and bovine serum albumin lie in the same range of 1-8 hr. The values of τ_2 thus appear to be dependent on intermolecular interactions in the monolayers rather than on the flexibility of individual molecules.

The adsorption of soluble substances at interfaces can usually be described in terms of the Gibbs adsorption isotherm

$$d\gamma = -RT \sum_{i} \Gamma_i \, d \ln a_i \tag{2.21}$$

where the summation extends over all species i in the system, and Γ_1 and a_i are the surface excess and activity in solution, respectively. In terms of surface pressure and area per molecule at the interface (A), we may write

$$d\pi = \frac{kT}{A} \, d \ln C_p \tag{2.22}$$

assuming protein activity is equal to C_p.

Thus the plot of π vs. ln C_p should be linear and can be used to get a value of A. Attempts to describe the equilibrium adsorption of globular proteins at the air–water interface in these terms have not been satisfactory in that the values of A so derived bear no relation to the size of the molecule and range from 1.58 nm^2 for bovine serum albumin (mol. wt. 66,000) to 6.72 nm^2 for pepsin (mol. wt. 35,000). Use of the above equations requires that the adsorption process is reversible, however, and there is clear evidence that this is not usually the case for protein monolayers; although desorption from protein monolayers can occur under certain conditions (MacRitchie, 1985), the processes occurring during desorption are not the reverse of the processes occurring during adsorption.

2.8.2.2. Structure of surface layers. Establishing the conformation of proteins in surface layers is a difficult problem. The spectroscopic techniques that have been applied so effectively to polypeptide monolayers (sec. 2.7.1) have not been used to any extent to investigate protein monolayers. Dilute films for which the surface concentration is 1 mg m^{-2} and the surface pressure is ≤ 0.1 mN m^{-1} are of the gaseous type and should obey the classical two-dimensional form of the equation of state

$$\pi A = kT \tag{2.23}$$

where A is the area per molecule.

The structure of concentrated films is more complex, particularly for molecules with no well-defined secondary or tertiary conformation such as β-casein. There have been numerous attempts to derive an equation of state for a polymer monolayer in order to relate the surface pressure to the conformation of the molecule in the adsorbed layer. The Singer (1948) equation applies to a flexible macromolecule of x statistical segments that are all adsorbed flat at the interface.

$$\pi = \pi_0 \left[\left(\frac{x-1}{x} \right) \frac{z^1}{2} \ln \left(\frac{1 - 2\theta}{z^1} \right) - \ln (1 - \theta) \right] \tag{2.24}$$

where π_0 is kT/a_0; $a_0 = $ the limiting area per segment; $\theta = $ degree of surface coverage ($= a_0/a$); and $z^1 = $ lattice coordination number, which for a flexible molecule is ~ 4 or for a rigid molecule is ~ 2.

Only two parameters (a_0 and z^1) are required to fit π-A curves on the basis of this equation, since for a high molecular weight material $x \gg 1$. A more sophisticated approach to the problem must take into account loop/tail formation as well as segment–segment and segment–solvent interactions. The simplest equation that allows for loop/tail conformations is that of Frisch and Simha (1956)

$$\pi = \pi_0 \left[\frac{S}{2} \frac{(\nu - 1)}{\nu} \ln \left(1 - \frac{2p\theta}{S} \right) - \ln (1 - p\theta) \right] \tag{2.25}$$

where π_o is kT/a_o; a_o is the limiting area per segment; ν is number of segments in the interface (i.e., in trains); p is fraction of the total number (t) of segments in the interface (i.e., ν/t); θ is a_o/a; and S is a number approximately equal to the lattice coordination number.

Equation 2.24 goes to the Singer equation when p = 1. More parameters $(a_o, p, \nu, $ and S) are required to fit a π-A curve for a flexible polymer with this equation.

The Singer equation can sometimes be fitted to protein π-A isotherms at low surface pressures. Thus the π-A isotherm of β-casein follows the Singer equation with $z^1 = 4$ and $a_o = 0.15$ nm^2 per amino acid residue at surface pressures < 8 mN m^{-1}, implying that the molecule is lying flat at the interface, but a rigid protein such as lysozyme does not obey the Singer equation even at a low surface pressure.

A comparison of the sequence of events in the adsorbed films of β-casein and lysozyme as suggested by Graham and Phillips (1979) is shown in Figure 2.18. These two proteins are probably representative of extremes of behavior from high flexibility in the case of β-casein to rigidity in the case of lysozyme. For β-casein, as the surface pressure and concentration increase, loop formation increases, and at saturation, multilayers are formed. In contrast to β-casein,

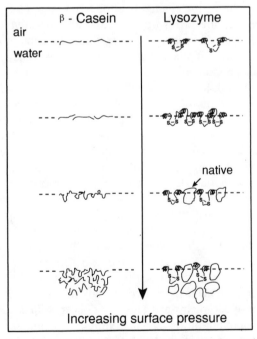

Fig. 2.18. Schematic representation of the structure of proteins at the air–water interface as the surface pressure (and surface concentration) are increased. From Graham and Phillips (1979), with permission of the publisher.

even at low surface pressures the lysozyme molecules are not fully unfolded and parts of the secondary structure (α-helices) remain. As the surface pressure and concentration are increased, there comes a point at which molecules reaching the interface cannot unfold against the prevailing surface pressure, and hence both unfolded and native molecules coexist.

2.8.2.3. Protein-lipid interactions in monolayers.

The interactions between proteins and lipid monolayers—particularly phospholipid monolayers—are of importance to both the structure and function of membranes. There are numerous examples of protein-lipid interactions that have been investigated by monolayer techniques—e.g., prothrombin-phosphatidylserine in relation to blood coagulation (Kop et al., 1984), cholora toxin–ganglioside G_{M1} interaction (Reed et al., 1987), colicin A (a bactericidal protein which depolarizes bacterial membranes)–phospholipid interaction (Frenette et al., 1989), and β-lactoglobulin-phospholipid interaction in relation to the stabilization of fat globules in milk (Cornell and Patterson, 1989). A change in surface pressure resulting from the injection of a protein into the monolayer substrate is taken as indicative of an interaction between the monolayer and the protein. If there is a strong affinity between the protein and the monolayer, the surface pressure will rise due to the penetration of the protein into the monolayer. The extent of penetration generally depends inversely on the initial surface pressure (π_i) of the monolayer, and a plot of the change in surface pressure ($\Delta\pi$) as a function of the initial surface pressure (π_i) is often linear with a negative slope and can be extrapolated to a value of π_i, at and above which the protein cannot penetrate the monolayer. Clearly, if the interaction between monolayer and protein is strong, the surface pressure above which it appears not to interact and penetrate will be high. In general, all proteins will interact with phospholipid monolayers under appropriate conditions, which can be found by manipulation of pH and/ or ionic strength. Positive proteins will bind electrostatically to negative phospholipids and vice-versa. Monolayer studies can be used to investigate specificity of a protein for particular phospholipids by addition of increasing amounts of the phospholipid of interest to a mixed monolayer with an inert phospholipid.

Miscibility, an important factor in protein-lipid monolayers, can be investigated by transferring the mixed monolayer from the air–water interface to a solid substrate for examination by electron microscopy. Phospholipids that give condensed monolayers have been found not to mix homogeneously with proteins such as β-lactoglobulin, bovine serum albumin, or β-casein, whereas homogeneous films were found when the phospholipid was forming on expanded film at the air–water interface (Cornell and Carroll, 1985).

The monolayer approach can be used to study the interactions between membrane proteins or glycoproteins and particular phospholipids. For example, the major sialoglycoloprotein of the human erythrocyte membrane, glycophorin A, is easily isolated in a pure form, and its interaction with monolayers has been studied (Davies et al., 1984). Glycophorin A (mol. wt. 31,000) contains approximately 60% carbohydrate and once isolated is water-soluble. It has 131 amino acid residues and three structural domains. The N-terminal domain (residues 1–71) is glycosylated and carries 16 oligosaccharide chains terminating in sialic acid residues that are largely responsible for the surface charge on the red blood cell. The transmembrane domain (residues 71–92) consists of 22 mainly hydrophobic amino acid residues that cross the erythrocyte bilayer in the α-helical conformation. The cytoplasmic C-terminal domain (residues 93–131) contains a large number of imino and acidic amino acids. Glycophorin can be spread at the aqueous–air interface provided the salt concentration in the substrate is high (e.g., 1.6 M ammonium sulphate); it will not, however, appreciably penetrate a phospholipid monolayer unless the surface pressure is low (below 2 mN m^{-1}). Mixed glycophorin-phospholipid monolayer can be obtained provided the glycophorin is spread first, followed by the phospholipid. For a phospholipid such as dipalmitoylphosphatidylcholine (DPPC)—which exhibits a characteristic LE to LC monolayer transition associated with the acyl chains, passing from an expanded (liquid-crystalline) to a condensed (gel state) monolayer on compression—at glycophorin to DPPC molecular ratios in the approximate range 0.003 to 0.025, the LE to LC transition is lost due to the interaction. Figure 2.19 shows π-A isotherms of mixed monolayers of glycophorin and DPPC in comparison with those calculated by simple additivity according to the equation.

$$A(\text{area per lipid})_{\text{at } \pi} = \frac{1}{N_L} (N_G A_G + N_L A_L) \qquad (2.26)$$

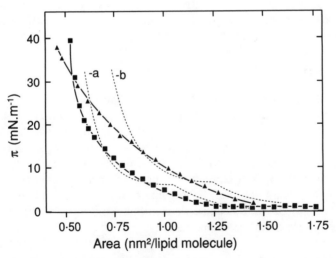

Fig. 2.19. Surface pressure (π) as a function of area per lipid molecule (A) for mixed monolayers of glycophorin and L-α-dipalmitoylphosphatidylcholine at molar ratios of 1:355 (■) and 1:178 (▲) on 1.6 M ammonium sulphate, pH 7.0 at 20°C. The initial surface concentration of glycophorin was 0.083 mg m^{-2}. The dotted curves a and b were calculated from the π-A isotherms of the pure components, assuming additivity. From Davies et al. (1984), with permission of the publisher.

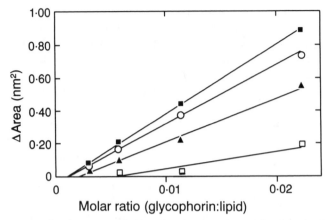

Fig. 2.20. Contraction in area per lipid molecule (ΔA) due to glycophorin-lipid (DPPC) interaction as a function of glycophorin to lipid molar ratio. ■, $\pi = 30$ mNm^{-1}; ○, $\pi = 25$ mNm^{-1}; ▲, $\pi = 20$ mNm^{-1}; □, 15 mNm^{-1}. From Davies et al. (1984), with permission of the publisher.

where N_G and N_L are the number of glycophorin and lipid molecules having areas per molecule A_G and A_L at a surface pressure π. Depending on the glycophorin to lipid ratio between surface pressures of 15 to 30 mN m^{-1}, the area per lipid molecule decreases relative to that calculated for a noninteracting mixture. However, at very low glycophorin to lipid molar ratios, the phase transition can still be observed. These observations suggest that there is a critical ratio of glycophorin to lipid below which excess lipid exists which can undergo chain melting. This critical ratio can be found by plotting the contraction in area (ΔA) as a function of molar ratio of glycophorin to lipid and extrapolating to zero ΔA. Figure 2.20 shows such a plot, from which it is deduced that the number of lipids that glycophorin can withdraw from participating in chain melting at the interface is 1,270 (at 25–30 mN m^{-1}), 460 (at 20 mN m^{-1}), and 210 (at 15 mN m^{-1}). Thus at the aqueous-air interface glycophorin is capable of influencing the behavior of large numbers of lipid molecules. These results are similar to those found for the numbers of DPPC molecules withdrawn from participation in chain melting, obtained by differential scanning calorimetry of DPPC liposomes incorporating glycophorin, where 197 (± 28) DPPC molecules were withdrawn from participation in chain melting per glycophorin molecule.

2.9. EQUIVALENCE OF MONOLAYERS AND BILAYERS

To be most useful as a model of behavior in a natural membrane, the monolayer must be compared to the bilayer at a surface pressure at which it most closely resembles the bilayer state. There have been several discussions on the problem of defining a monolayer surface pressure that corresponds to bilayer behavior (Ohki and Ohki, 1976; Nagle, 1976; Marcelja, 1974). Some of the problems relating to the meaning of surface pressure have been pointed out by Nagle (1986). In the case of the monolayer ($\pi = \gamma_o - \gamma$), the surface pressure is the pressure that must be applied to the monolayer to hold it at a given area per molecule. In the case of a bilayer, no external pressure must be applied, so that by the above definition $\pi = o$. But monolayers at $\pi = o$ do not correspond to bilayers (as Gorter and Grendel assumed) for two reasons. First, as defined above, π of the monolayer involves the surface tension of water, which should be added to the surface pressure to get the total monolayer tension ($\pi + 72$) mN m^{-1}. Second, the monolayer is at the air–water interface, not the oil–water interface to which the bilayer may be approximated. If we consider a film of oil on a water substrate, then at the intersection of the three interfaces, oil–water–air (o–w–a), assuming zero contact angle

$$\gamma_{a/w} = \gamma_{a/o} + \gamma_{o/w} \qquad (2.27)$$

so that the difference in tension between an air–water ($\gamma_{a/w}$) and an oil–water ($\gamma_{o/w}$) surface is $\gamma_{a/o}$, which is of the order of 20 mN m^{-1}. Thus correcting for both of these effects suggests that the monolayer at a surface pressure of approximately 50 mN m^{-1} should be equivalent to the bilayer state. On the basis of a comparison of the absolute values of the molecular area of phospholipid molecules in the bilayer gel phase and the change in area at the bilayer and monolayer phase transitions, Blume (1979) concluded that the bilayer system is very similar to that of the corresponding monolayer system at a surface pressure of approximately 30 mN m^{-1}, because at this pressure the absolute area and the change in area in both systems are identical. From these considerations we can conclude that monolayers model bilayer behavior at surface pressures in the region of 30–50 mN m^{-1}, with the latter figure being more likely.

Another parameter that has been used to compare model membrane systems such as monolayers and liposomes with biological membranes is "internal pressure." Measurements of the partitioning of amphipathic molecules such as chlorpromazine and decanol between cell membranes and the aqueous phase in comparison with partitioning into liposome bilayers showed that the solubility of these amphiphiles was very much lower in natural membranes than in the model systems (Conrad and Singer, 1979). On this evidence it was suggested that natural membranes had a much higher internal pressure, although the use of the term *internal pressure* was meant to be taken figuratively rather than literally (Conrad and Singer, 1981). It was suggested that the higher "internal pressure" of natural membranes arose from the presence of integral membrane proteins. Evidence against the existence of a protein-induced high internal pressure was found from the comparison of the partitioning of 8-anilinonaphthalene-1-sulphate into Triton X-100 micelles and submitochondrial particles; partitioning was found to be very similar in both cases (Gaines and Dawson, 1982).

Internal pressure (P_i) can be defined rigorously in thermodynamic terms and in two dimensions is given by (Davies and Jones, 1991).

$$P_i = T \left(\frac{d\pi}{dT}\right)_A - \pi \qquad (2.28)$$

From equation 2.28 it is possible to calculate P_i from the temperature coefficient of the surface pressure of a monolayer. Using this method, it was found that under conditions in which a monolayer is believed to most closely model a bilayer, the internal pressures of pure dioleoylphosphatidylcholine (DOPC) and DPPC monolayers were 0.32 Nm^{-1} and 0.13 Nm^{-1}, respectively, while for a mixed glycophorin-DPPC monolayer, P_i was 0.36 Nm^{-1}. These measurements give no support for the view that integral membrane proteins lead to abnormally high internal pressures in membranes.

REFERENCES AND FURTHER READING

Adam, N.K. [1941] (1968) The Physics and Chemistry of Surfaces. Dover, New York, pp. 17–105.

Aveyard, R. and Haydon, D.A. (1973) An Introduction to the Principles of Surface Chemistry. Cambridge University Press, Cambridge.

Baglioni, P., Cestelli, G., Dei, L. and Gabrielli, G. (1985) Monolayers of cholesterol at water–air interface: Mechanism of collapse. J. Coll. Int. Sci. *104*, 143–150.

Bar, R.S., Deamer, D.W. and Cornwell, D.G. (1966) Surface area of human erythrocyte lipids: Reinvestigation of experiments on plasma membranes. Science *153*, 1010–1012.

Bayerl, T.M., Thomas, R.K., Penfold, J., Rennie, A. and Sackmann, E. (1990) Specular reflection of neutrons at phospholipid monolayers. Changes of monolayer structure and headgroup hydration at the transition from the expanded to the condensed phase state. Biophys. J. *57*, 1095–1098.

Blume, A. (1979) A comparative study of the phase transitions of phospholipid bilayers and monolayers. Biochim. Biophys. Acta *557*, 32–44.

Boonman, A., Machiels, F.H.J., Snik, A.F.M. and Egberts, J. (1987) Squeeze-out from mixed monolayers of dipalmitoylphosphatidylcholine and egg phosphatidylglycerol. J. Coll. Int. Sci. *120*, 456–468.

Clements, J.A. (1977) Functions of the alveolar lining. Am. Rev. Respir. Dis. *115*, 67–71.

Conrad, M.J. and Singer, S.J (1979) Evidence for a large internal pressure in biological membranes. Proc. Natl. Acad. Sci. USA *76*, 5202–5206.

Conrad, M.J. and Singer, S.J. (1981) The solubility of amphipathic molecules in biological membranes and lipid bilayers and its implications for membrane structure. Biochemistry *20*, 808–818.

Cordoba, J., Jackson, S.M. and Jones, M.N. (1990) Mixed monolayers of phosphatidylinositol and dipalmitoylphosphatidylcholine and their interaction with liposomes. Colloids Surf. *46*, 85–94.

Cornell, D.G. and Carroll, R.J. (1985) Miscibility in lipid-protein monolayers. J. Coll. Int. Sci. *108*, 226–233.

Cornell, D.G. and Patterson, D.L. (1989) Interaction of phospholipids in monolayers with β-lactoglobulin absorbed from solution. J. Agric. Food Chem. *37*, 1455–1459.

Davies, J.T. and Rideal, E.K. (1963) Interfacial Phenomena. 2nd ed. Academic Press, New York, pp. 217–281.

Davies, R.J. and Jones, M.N. (1992) The thermal behaviour of phosphatidylcholine–glycophorin monolayers in relation to monolayer and bilayer internal pressure. Biochim. Biophys. Acta *1103*, 8–12.

Davies, R.J., Goodwin, G.C., Lyle, I.G. and Jones, M.N. (1984) The interaction of glycophorin with dipalmitoylphosphatidylcholine at the air–water interface. Colloids Surf. *8*, 261–270.

Dervichian, D. and Macheboeuf, M. (1938) Sur l'existence d'une couche monomoléculaire de substances lipoidiques a'la surface des globules rouges du song. Compt. Rend. *206*, 1511–1514.

Fontanges, A. de, Bonté, F., Taupin, C. and Ober, R. (1984). Pressure-area curves of single and mixed monolayers of phospholipids and the possible relevance to properties of lung surfactant. J. Coll. Int. Sci. *101*, 301–308.

Frenette, M., Knibiehler, M., Baty, D., Geli, V., Pattus, F., Verger, R. and Lazdunski, C. (1989) Interactions of colicin A domains with phospholipid monolayers and liposomes: Relevance to the mechanism of action. Biochemistry *28*, 2509–2514.

Frisch, H.L. and Simha, R. (1956). Monolayers of linear macromolecules. J. Chem. Phys. *24*, 652–655.

Gabrielli, G., Baglioni, P. and Ferroni, E. (1981) On the mechanism of collapse of monolayers of macromolecular substances: Poly (L, D and DL) alanine. J. Coll. Int. Sci. *81*, 139–149.

Gains, G.L. (1966) Insoluble Monolayers at Liquid Gas Interfaces. Interscience, New York.

Gaines, N. and Dawson, A.P. (1982) Evidence against protein-induced "internal pressure" in biological membranes. Biochem. J. *207*, 567–572.

Goodrich, F.C. (1957) in Proceedings of the Second International Congress of Surface Activity. Vol. 1. J.H. Schulaman, Ed. Butterworths Publishers, London, pp. 85–91.

Gorter, E. and Grendel, F. (1925) On biomolecular layers of lipids on the chromocytes of the blood. J. Exp. Med. *41*, 439–443.

Graham, D.E. and Phillips, M.C. (1979) Proteins at interfaces. Part 1: Kinetics of adsorption and surface denaturation. Part 2: Adsorption isotherms. J. Coll. Int. Sci. *70*, 415–426; 427–439.

Heckl, W.H., Cadenhead, D.A. and Möhwald, H. (1988) Cholesterol concentration dependence of quasi-crystalline domains in mixed monolayers of the cholesterol-dimyristoylphospatidic acid system. Langmuir *4*, 1352–1358.

Helm, C.A., Laxhuber, L.A., Lösche, M. and Möhwald, H. (1986) Electrostatic interactions in phospholipid membranes. Part 1: Influence of monovalent ions. Colloid Polymer Sci. *264*, 46–55.

Knobler, C.M. (1990a) Recent developments in the study of monolayers at the air-water interface. Adv. Chem. Phys. (Ed. I. Prigogine and S.A. Rice) *77*, 397–449.

Knobler, C.M. (1990b) Seeing phenomena in flat land: Studies of monolayers by fluorescence microscopy. Science *249*, 870–874.

Kop, J.M.M., Cuypers, P.A., Lindhout, T., Hemker, H.C. and Hermens, W.T. (1984) The adsorption of prothrombin to phospholipid monolayers quantitated by ellipsometry. J. Biol. Chem. *259*, 13993–13998.

Lee, E.M., Thomas, R.K., Penfold, J. and Ward, R.C. (1989) Structure of aqueous decyltrimethylammonium bromide solutions at the air/water interface studied by the specular reflection of neutrons. J. Phys. Chem. *93*, 381–388.

Lösche, M. and Möhwald, H. (1989) Electrostatic interactions in phospholipid membranes. Part 2: Influence of divalent ions on monolayer structure. J. Coll. Int. Sci. *131*, 56–66.

MacRitchie, F. (1985) Desorption of proteins from the air/water interface. J. Coll. Int. Sci. *105*, 119–123.

MacRitchie, F. (1990) Chemistry at Interfaces. Academic Press, London.

MacRitchie, F. and Alexander, A.E. (1963) Kinetics pf adsorption of proteins at interfaces. Part 1–3. J. Coll. Int. Sci. *18*, 453–458; 458–464; 464–469.

Malcolm, B.R. (1973) in Progress in Surface and Membrane Science. Vol. 7. J.F. Danielli, M.O. Rosenberg, and D.A. Cadenhead, Eds. Academic Press, New York, pp. 183–229.

Marcelja, S. (1974) Chain ordering in liquid crystals. Part 2: Structure of bilayer membranes. Biochim. Biophys. Acta *367*, 165–176.

Morley, C.J. and Bangham, A.D. (1981) Use of surfactant to prevent respiratory distress syndrome. Prog. Respir. Res. *15*, 261–268.

Nagle, J.F. (1976) Theory of lipid monolayer and bilayer phase transitions: Effect of headgroups interactions. J. Membr. Biol. *27*, 233–250.

Nagle, J.F. (1986) Theory of lipid monolayer and bilayer chain-melting transitions. Discuss. Faraday Soc. *81*, 151–162.

Ohki, S. and Ohki, C.B. (1976) Monolayers at the oil–water interface as a proper model for bilayer membranes. J. Theor. Biol. *62*, 389–407.

Pallas, N.R. and Pethica, B.A. (1985) Liquid-expanded to liquid-condensed transitions in lipid monolayers at the air/water interface. Langmuir *1*, 509–513.

Papahadjopoulos, D. (1968) Surface properties of acidic phospholipids: Interactions of monolayers and hydrated liquid crystals with uni- and bi-valent ions. Biochim. Biophys. Acta *163*, 240–254.

Phillips, M.C. and Chapman, D. (1968) Monolayer characteristics of saturated 1,2-diacylphosphatidylcholines (lecithins) and phosphatidylethanolamines at the air-water interface. Biochim. Biophys. Acta *163*, 301.

Phillips, M.C., Jones, M.N., Patrick, C.P., Jones, N.B. and Rodgers, M. (1979) Monolayers of block copolypeptides at the air–water interface. J. Coll. Int. Sci. *72*, 98–105.

Reed, R.A., Mattai, J. and Shipley, G.G. (1987) Interaction of cholera toxin with ganglioside G_{M1} receptors in supported lipid monolayers. Biochemistry *26*, 824–832.

Singer, S.J. (1948) Note on an equation for linear macromolecules in monolayers. J. Chem. Phys. *16*, 872–876.

Tajima, K. and Gershfeld, N.L. (1985) Phospholipid surface bilayers at the air–water interface. Part 1: Thermodynamic properties. Biophys. J. *47*, 203–209.

Tenenbaum, A., Berliner, C., Ruysschert, J.M. and Jaffe, J. (1976) Conformation of oligopeptides at the air–water interface: A fluorometric study. J. Coll. Int. Sci. *56*, 360–364.

Ter-Minassian–Saraga (1981) Protein denaturation on adsorption and water activity at interfaces: An analysis and suggestion. J. Coll. Int. Sci. *80*, 393–401.

Ward, A.F.H. and Tordai, L. (1946) Time-dependence of boundary tensions of solutions. Part 1: Role of diffusion in time-effects. J. Chem. Phys. *14*, 453–461.

CHAPTER 3

THE MICELLAR STATE

3.1. THE CRITICAL MICELLE CONCENTRATION AND MICELLE FORMATION

The chemistry of surface-active materials, including synthetic surfactants, natural surfactants such as the bile salts, and membrane lipids, has been discussed in Chapter 1. A characteristic feature of synthetic surfactants and bile salts in aqueous solutions that is a direct consequence of their amphipathic nature is the formation of aggregates or micelles. Micelle formations usually occur over a very narrow concentration range as the total surfactant concentration is raised. It is often a highly cooperative process and is characterized by abrupt changes in a whole variety of solution properties; the most frequently used properties to define the critical micelle concentrations (cmc) are surface tension, conductivity, and turbidity (or light scattering). Figure 3.1 shows schematic plots of these properties as a function of the total surfactant concentration in solution. The cmc is defined by the sharp changes in these curves and corresponds to the point at which micelles first form in the solution. The cmc may differ slightly, depending on the nature of the physical technique used to determine it. This is because different physical techniques respond differently to monomeric and micellar species in the solution, but such differences are not usually large unless the particular surfactant forms small micelles with a low degree of cooperativity. The number of surfactant monomers per micelle—i.e., the aggregation number—varies from less than 10 to 100 or more, and the larger the aggregation number the sharper the cmc (see below).

The behavior of ionic surfactants at concentrations below the cmc appears to differ little from that of a simple electrolyte. The conductivity of such solutions increases linearly with surfactant concentration and can be interpreted

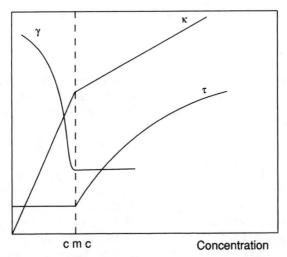

Fig. 3.1. Schematic plots of surface tension (γ), specific conductivity (κ) and turbidity (τ) as a function of concentration for a typical surfactant. The dotted line denotes the critical micelle concentration (cmc).

in terms of the Onsager theory of electrolyte conductivity for anionic surfactants such as sodium n-dodecylsulphate (SDS) (Parfitt and Smith, 1962); although some workers have produced evidence for the existence of pre-cmc association and suggest that dimeric species can exist below the cmc by association of two alkyl chains of the surfactant (van Voorst Vader, 1961), the evidence for the occurrence of such pre-cmc association is not strong. At the cmc, the rate of increase of the conductivity of the solution with concentration decreases, i.e., in terms of specific conductivity (κ).

$$\left(\frac{d\kappa}{dc}\right)_{c<cmc} > \left(\frac{d\kappa}{dc}\right)_{c>cmc} \tag{3.1}$$

Considering the effect of micellization on solution conductivity, if it is assumed that an approximately spherical micelle forms from N monomers, the volume (V) of the micelle of radius R would be simply $(4/3)\pi R^3$ and hence R is proportional to $\sqrt[3]{V}$ and hence to the cube root of the number of monomers N (i.e., $R \propto \sqrt[3]{N}$). The frictional resistance of the micelle would thus depend on $\sqrt[3]{N}$ through Stokes Law, $6\pi\eta Rv$, where η is the viscosity of the solvent and v the micellar velocity. The electrical force on the micelle for a singly charged monomer will depend directly on its charge and hence directly on the aggregation number N and be equal to EN, where E is the electrical field strength. Thus the electrical force on the micelle increases directly with N, while the frictional resistance increases only as $\sqrt[3]{N}$. One would hence expect that on micellization the specific conductivity should increase more rapidly with concentration than for an equivalent number of monomers. Experimentally, however, this is not the case at low field strengths, and the reason for this is that although ionic surfactants are strong electrolytes below the cmc (i.e., fully ionized), the charge density on the micellar surface is so high that counterions (ions of opposite charge to the micelle) are bound to the surface and reduce the net charge, so that the degree of ionization is often only 0.2 to 0.3 and hence only 20 to 30% of the surfactant monomers in the micelle are ionized.

There are two instances when the conductivity of micellar solutions increases more steeply with surfactant concentration than the increase observed for monomers. It was shown many years ago by Hartley (1936) that micelles in a very high electric field (~ 200 KV cm^{-1}) have an increased conductance due to the removal of the counterions from the ionic atmosphere around the micelle, so that they behave more as expected in the absence of counterion binding. The other instance is in the presence of high salt concentrations (>0.35 M) (Mysels and Mysels, 1965, 1972). In this case, however, the effect is due to a combination of a decrease in the conductance of the monomers with increasing salt and an increase in the effective ionization of the micelles with increasing salt.

The determination of the cmc from surface tension and turbidity depends on the adsorption of the surfactant monomer at an interface (air–aqueous or oil–aqueous) and the increased light scattering that accompanies the onset of micelle formation, respectively. In the case of surface or interfacial tension (Υ), the decrease with increasing monomeric surfactant concentration, [S], is described by the Gibbs Adsorption isotherm, which for an ideal solution may be written (Gibbs, 1931)

$$\frac{d\gamma}{d \ln [S]} = -x\Gamma_2 RT \tag{3.2}$$

where Γ_2 is the surface excess of surfactant at the interface (moles/unit area), R is the gas constant, and T the absolute temperature. The surface excess is the increase in the number of moles of surfactant per unit area of interface above that which would be there if the aqueous solution remained homogenous up to the interface. A detailed treatment of this fundamental equation can be found in numerous texts (e.g., Adamson, 1960; Jones, 1975). The parameter x is an integer that for an ionic surfactant in water has a value 2. For an ionic surfactant in the presence of a relatively high concentration of salt (e.g., 0.2 M), x is 1. It is also 1 for a nonionic surfactant. At intermediate salt concentrations (0.02 M), x varies between 2 and 1 (Bijsterbosch and Van den Hul, 1967). The decrease in γ with increasing surfactant concentration means that both $d\gamma/d$ [S] and $d\gamma/d \ln$ [S] are negative and Γ_2 is hence positive. The positive adsorption of a surfactant at an interface is one of the most fundamental properties of a surface-active material. After saturating the interface, the monomeric surfactant concentration continues to increase in the bulk solution until the cmc is reached. Above the cmc, the monomer concentration remains approximately constant, so that d ln [S] = 0; hence from equation 3.2 dγ = 0 and the surface tension remains approximately constant but importantly not exactly constant, because there is a dynamic equilibrium between the monomers and micelles above the cmc and the monomer "activity" increases very slowly above the cmc (see below), so that γ decreases very slightly (Elworthy and Mysels, 1966). Surface or interfacial tension plots can be used to determine both the surface excess and the cmc for both ionic and nonionic surfactants as can turbidity measurements, in contrast to conductivity measurements, which are generally only useful for ionic surfactants. Since turbidity depends on the size of the micelles, which depends on both the aggregation number and the extent of counterion binding, this technique enables these parameters to be determined as well as the cmc.

The picture we have of the micellization process is summarized in Figure 3.2 for a typical ionic surfactant. The details of the micellar structure will depend to some degree on the chemistry of the surfactant, as will the value of the cmc and the particular composition of the aqueous phase as discussed below.

3.1.1. The cmc of Synthetic Surfactants and the Effects of Composition of the Medium

The magnitude of the cmc of a surfactant is determined by the chemistry of the surfactant and the composition and temperature of the solvent. Table 3.1 shows the cmcs and other micellar parameters for a selection of synthetic surfactants. The overriding consideration with regard to the chemical structure of the surfactant and its cmc is the hydrophobic-hydropholic balance of the molecule. The tendency to form micelles is increased by increasing the hydro-

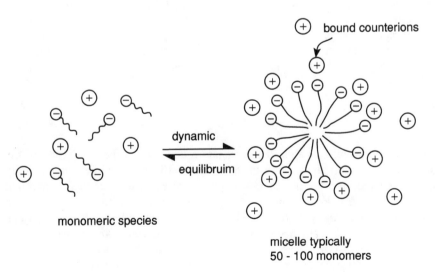

bound counterions

dynamic
equilibruim

monomeric species

micelle typically
50 - 100 monomers

Fig. 3.2. Schematic view of the micellization of a typical surfactant.

TABLE 3.1. Micellar Parameters for Typical Synthetic Surfactants in Aqueous Solutions

Surfactant	Medium/Temperature (°C)	cmc (mM)	N^a	p/N^a
Synthetic anionic				
Sodium n-octylsulphate[b]	H₂O/25°	130		
Sodium n-decylsulphate[b]	H₂O/25°	33		
Sodium n-dodecylsulphate[c]	H₂O/25°	8.1	58	0.18
Sodium n-dodecylsulphate[c]	0.1M NaCl/25°	1.4	91	0.12
Sodium n-dodecylsulphate[c]	0.2M NaCl/25°	0.83	105	0.14
Sodium n-dodecylsulphate[c]	0.4M NaCl/25°	0.52	129	0.13
Synthetic cationic				
n-Dodecyltrimethylammonium bromide[d]	H₂O/25°	14.8	43	0.17
n-Dodecyltrimethylammonium bromide[d]	0.0175M NaBr/25°	10.4	71	0.17
n-Dodecyltrimethylammonium bromide[d]	0.05M NaBr/25°	7.0	76	0.16
n-Dodecyltrimethylammonium bromide[d]	0.10M NaBr/25°	4.65	78	0.16
Synthetic nonionic				
PEG (4-5) p-t-octylphenol (TritonX-45)[e]	H₂O/25°	0.11		—
PEG (7-8) p-t-octylphenol (TritonX-114)[e]	H₂O/25°	0.20		—
PEG (9-10) p-t-octylphenol (TritonX-100)	H₂O/25°	0.24	140	—
n-Octyl β-D-glucopyranoside[f]	H₂O/25°	23	27	—

[a]N = aggregation number; p = micellar charge.
[b]Data from: Murkerjee and Mysels (1971).
[c]Kratohvil (1980).
[d]Jones and Piercy (1972).
[e]Helenius and Simons (1975).
[f]Wasylewski and Kozik (1979).

phobicity of the molecule and decreased by increase in hydrophilicity. Thus, for ionic surfactants with n-alkyl chains such as the n-alkyl sulphates, the cmcs decrease with increase in chain length. In general, for a homologous series of surfactants in a given solvent at a fixed temperature, there is a linear relationship between $\log_{10}$ cmc and the number of carbon atoms (n) in the chain

$$\log_{10} \text{cmc} = a - bn \qquad (3.3)$$

where a and b are constants. The constant a depends on the headgroup of the surfactant, typical values being in the range of 10 to 3.5, while b is generally of the order of 0.3 for an ionic surfactant and 0.5 for a nonionic surfactant (Kresheck, 1975). For the n-alkylsulphates in water, the cmcs decrease by a factor of approximately 4 on increasing the n-alkyl chain length by 2 methylene groups. In general, branched-chain and unsaturated hydrocarbon chain surfactants have lower cmcs than straight-chain surfactants, and the addition of a benzene ring into the chain is approximately equivalent to the introduction of 3.5 methylene groups.

Theoretical studies by Shinoda (1963) have shown that for an ionic uni-univalent $(1:1)$ surfactant, b is given by

$$b(1:1) = \left(\frac{1}{1 + K_g}\right)\frac{W}{2.303kT} \tag{3.4}$$

where K_g is related to the degree of counterion binding to the micelle and W is the hydrophobic interaction contribution to micelle formation, and k and T are the Boltzmann constant and absolute temperature, respectively. For a nonionic surfactant, $K_g = 0$ and hence b is given by

$$b(\text{nonionic}) = \frac{W}{2.303kT} \tag{3.5}$$

The values of W are of the order of $1.1 - 1.3$ kT, and K_g is of the order of 0.6 to 0.9.

The cmcs of ionic surfactants decrease on the addition of inert electrolytes to the solution. The increase in ionic strength due to addition of electrolytes effectively increases the hydrophobic interaction between the surfactant molecules and causes them to form micelles at a lower concentration than in water. There is a logarithmic relationship between the cmc and the concentration of counterions (c_i), which for a singly charged surfactant in a uni-univalent $(1:1)$ electrolyte may be written

$$\log_{10} \text{cmc} = -K_g \log c_i + C \tag{3.6}$$

As above, K_g is related to degree of counterion binding and may be specifically defined as the effective electrical work required to introduce another charge, e, into the micellar surface with a surface potential ψ_o, which is given by $K_g e \psi_o$. If the charge introduced is divalent, then the work done per unit charge would be $(K_g 2e\psi_o)/2$ and equation would become

$$\log_{10} \text{cmc} = -\frac{K_g}{2} \log c_i + C \tag{3.7}$$

Thus, for example, a series of tetradecyltrimethylammonium (TTMA) surfactants in $1:1$ electrolytes with a common counterion were found to have values of K_g of 0.663 (TTMA Cl + NaCl), 0.696 (TTMA Br + NaBr), and 0.666 (TTMANO$_3$ + NaNO$_3$), whereas for the sulphate salt (TTMA)$_2$SO$_4$ in Na$_2$SO$_4$, a $1:2$ electrolyte, the slope of $\log_{10}$ cmc vs. $\log_{10}$ [SO$_4$=] was found to be 0.375, giving a value of K_g of 0.75 in approximate agreement with the data for the $1:1$ electrolytes (Jones and Reed, 1968).

An important consideration with regard to the addition of electrolytes to surfactants is the solubility of the surfactant salts that may form. This is particularly important for anionic surfactants where the counterions are metallic. The divalent and trivalent metal salts of anionic surfactants are often insoluble, so that the addition of divalent ions which are commonly added to buffer systems in biological work—such as calcium and magnesium ions—readily precipitates n-alkylsulphates. Even for singly charged ions, solubility can be an important factor; for the n-alkylsulphates, the solubility of the alkali metal salts decreases in the sequence $Li^+ > Na^+ > Rb^+ > Cs^+$.

The temperature dependence of the solubility of surfactants shows a steep rise at a particular temperature known as the Krafft point. As the temperature of an aqueous surfactant solution increases, an increasing amount of "monomeric" surfactant dissolves. At the Krafft point the concentration of dissolved surfactant coincides with the cmc at that temperature, so that the surfactant now dissolves to form micelles and the rate of increase of solubility with temperature is markedly increased. The Krafft point for sodium n-dodecylsulphate in water is approximately $4\,^{\circ}C$, so that solutions containing this surfactant very readily precipitate in a laboratory refrigerator.

A further point relating to stability in surfactant solutions containing multivalent metallic ions is that in anionic surfactant solutions, these ions have a higher affinity for the surface of the micelle than do univalent ions, so that a small concentration of multivalent ions can be solubilized by adsorption to the

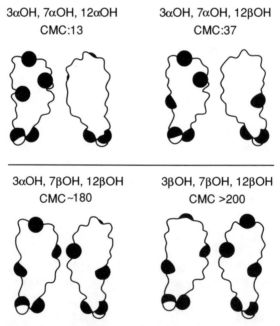

Fig. 3.3. Photographs of space-filling models of cholic acid (top left) and its three epimers. The hydroxyl group are the black circles and the carboxyl group can be seen at the base of each model. For each molecule the α-face is on the left and the β-face is on the right. From Roda et al. (1983), with permission of the publisher.

micellar surface above the cmc, but on dilution below the cmc salt of the surfactant monomer will precipitate from the solution.

Nonelectrolytes may also affect the cmc, the most commonly encountered nonelectrolytes being urea and alcohols. The extensive use of urea ($CO(NH_2)_2$) and the isoelectronically related compound guanidinuim chloride ($CNH(NH_2)_2H^+ Cl^-$) as protein denaturants relates to their property of weakening the hydrophobic interaction. Given this observation, it would be expected that urea would increase the cmc of a surfactant as a consequence of reducing the hydrophobic interaction between the apolar chains. In 6 M urea the cmcs of a wide range of surfactant types are increased by factors of between 1.25 and 4.0 (Kresheck, 1975), consistent with a weakening of hydrophobic interaction. Thus equations analogous to equations 3.4 and 3.5 above could be written in which W would have a lower value in urea than in water. Urea may also affect headgroup hydration to some degree.

The addition of organic molecules such as dioxane or ethanol to aqueous surfactant solutions will in general raise the cmc of a surfactant by making the aqueous phase more hydrophobic. Thus addition of 25 wt% dioxane to aqueous sodium n-dodecylsulphate raises the cmc at 25°C from 8 mM to 30 mM; likewise 25 wt% of ethanol raises the cmc to 10.7 mM.

3.1.2. The cmc of Natural Surfactants

The largest group of natural surfactants are the bile acids, which on conjugation with glycine ($NH_2 CH_2 CO_2H$) and taurine ($NH_2 CH_2 CH_2 SO_3H$) give the bile salts (see Chap. 1, sec. 1.1.2). Both the bile acids and their salts associate in aqueous solution, and a cmc can be measured by surface tension or dye solubilization. In the latter method, the amount of dye solubilized is measured spectroscopically as a function of surfactant concentration. The absorbance increases sharply at the cmc, when the dye is taken up by the micelles. Orange OT, azulene, and rhodamine 6G have been used to determine the cmc of bile acids and salts. The formation of micelles by bile acids and salts occurs over a fairly narrow concentration range, and the existence of a true cmc has been questioned. It has been suggested that the term *uncritical multimer concentration* would more suitably describe the behavior (Hofmann and Mysels, 1988). However, the term *cmc* is still in general use and the various methods used to measure the cmc give consistent values suggesting there is not too much ambiguity in defining a cmc.

The cmcs of a large number of bile compounds in water and salt solution have been collected and carefully assessed by Roda et al. (1983), and a selection of values for the predominant human biliary compounds is shown in Table 3.2. The cmcs of the bile compounds in general range from 1 to > 250 mM and vary subtly with the stereochemistry of the molecules. For example, changing the configuration of the hydroxy group in position 7 from α to β in the dihydroxy acids chenodeoxycholic and ursodeoxycholic increases the cmc by a factor of more than 2 (Table 3.2).

From a study of space-filling models of the bile acids, the hypothesis that

TABLE 3.2. Critical Micelle Concentrations of the Bile Compounds in Aqueous Solution

Compound	Hydroxy Substituent 3	7	12	Unconjugated Acid cmc (H$_2$O) mM	cmc (NaCl)[a] mM	Bile Salt Glycine cmc (H$_2$O) mM	cmc (NaCl)[a] mM	Taurine cmc (H$_2$O) mM	cmc (NaCl) mM
Cholic acid	α-OH	α-OH	α-OH	13	11	12	10	10	6
Chenodeoxycholic	α-OH	α-OH		9	4	6	1.8	7	3
Ursdeoxycholic	α-OH	β-OH		19	7	12	4	8	2.2
Deoxycholic	α-OH		α-OH	10	3	6	2	6	2.4

From Roda et al. (1983), with permission of the publisher.

[a]0.15 M sodium chloride.

the cmc is dependent on the size of the contiguous hydrophobic area of the molecule was proposed, and the greater this area the lower the cmc. Figure 3.3 shows the shape and position of the three hydroxyl and carboxyl groups for cholic acid and its three epimers, together with the cmcs. The β-faces of the molecules are shown on the right of each pair. Thus for cholic acid (cmc 13 mM) the three hydroxy groups are all α and the β-face of the molecule has a large contiguous hydrophobic area; on changing the configuration of the 12-OH from α to β, the β-face has an OH group in what was previously a hydrophobic area and the cmc increases to 37 mM. A further increase in cmc occurs when the 7-OH configuration is changed from α to β, and when all three OH groups are β, both sides of the molecule are hydrophilic and the cmc exceeds 200 mM. In general, conjugation of the bile acids with glycine or taurine does not markedly change their cmcs, but like all anionic surfactants, addition of salt decreases the cmc.

The acidity of aqueous solutions has a marked effect on the solubility of bile acids. The pK_a of the carboxyl groups in bile acids is of the order of 5, so that if the pH is appreciably below 5, protonation of the carboxyl group will cause the acid to become insoluble and crystallize from solution. There is a relatively narrow pH range above which the solubility of bile acids increases sharply as the acid dissolves in micellar form. This has been called the critical micellar pH (CmpH), which, like the cmc, depends on the size of the contiguous area on the hydrophobic side of the molecule. Figure 3.4 shows the pH dependence

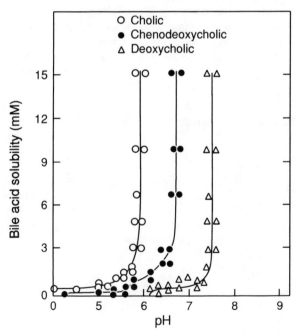

Fig. 3.4. Solubility of bile acids as a function of pH. From Hofman and Mysels (1988), with permission of the publisher.

of the solubility of the three most common bile acids. Conjugation of the bile acids significantly changes the pK_a; thus amidation with glycine lowers the pK_a to approximately 3.9 and for the taurine conjugates to < 1. The conjugated bile acids are thus more soluble at lower pH than the unconjugated forms. This has important physiological consequences; the pH in the small intestine is between 5.5 and 7 so that the bile salts will remain soluble.

3.1.3. The cmc of Phospholipids

As major components of cell membranes, the properties of phospholipids in aqueous systems are of great importance. Looking at phospholipid structures from the point of view of surfactant chemistry, the double fatty acyl chain in conjunction with a singly charged or zwitterionic headgroup suggests that the hydrophobic-hydrophilic balance of the molecules lies toward hydrophobicity, and this is certainly the case for the long-chain phospholipids found in membranes. As will be discussed in Chapter 4, they form a variety of lytropic mesophases in aqueous media, the structures of which are largely based on the bilayer arrangement of the molecules, with the fatty acyl chains back-to-back and the headgroups satisfying their affinity for water. We will discuss the question of the existence of monomeric long-chain phospholipids in aqueous media later, but it is interesting to consider the properties of the shorter-chain analogues, which essentially behave like surfactants with well-defined cmcs above which micelles are formed.

A thorough study of the short-chain phosphatidylcholines has been made and their cmcs have been measured by several techniques, including surface tension and light scattering. Table 3.3 summarizes a selection of values for short-chain lecithins in water. The cmcs decrease with increasing chain length, so that for the diC_9 lecithin, the cmc is in the micromolar range. These short-chain lecithins are surface-active, the surface tension at the air–aqueous interface falling to 22 mN m^{-1} at the cmc for the diC_9 lecithin. On addition of electrolytes, the cmcs are lowered; the dependence of cmc on salt concentration

TABLE 3.3. Critical Micelle Concentrations of Short-Chain Lecithins in Aqueous Media

Lecithin (Phosphatidylcholine)	Formula Mol. Wt. as Monohydrate	Surface Tension at the cmc (mM)	cmc (mM)
Dibutanoyl lecithin (diC_4)	416.4		80 ± 5[a]
Dihexanoyl lecithin (diC_6)	472.5	30.6	14.6
Diheptanoyl lecithin (diC_7)	500.6	26.6	1.42
Dioctanoyl lecithin (diC_8)	528.7	24.0	0.265
Dinonanoyl lecithin (diC_9)	556.7	22.0	0.00287

[a]At 45°C (Wells, 1974). The other data relate to room temperature measurements (20–25°C) (Tausk et al. [1974a]).

follows an equation applicable to nonionic or zwitterionic surfactants, which may be written (compare equation 3.6)

$$\log \text{cmc} = -k_s c_i + (\log \text{cmc})_{c_i = 0} \tag{3.8}$$

where k_s may be identified as a specific salt-effect constant related to the ability of a given salt to "salt out" a nonpolar solute from solution. For the diC_6 lecithin, the k_s values vary with salt as follows: $1.04\,(Na_2SO_4)$, $0.6\,(NaF)$, $0.26\,(NaCl)$, and $0.05\,(LiI)$, which is approximately in accordance with the lytopic series for the effectiveness of anions to salt out lyophilic colloids, e.g., proteins. That is,

$$\tfrac{1}{2}SO_4^{2-} > F^- > Cl^- > ClO_4^- > Br^- > NO_3^- > I^- > SCN^-$$

It should be noted that the depression of the cmc by salting-out occurs only at relatively high salt concentrations. For example, for the diC_6 lecithin with a cmc in water of 14.6 mM, the cmc decreases by approximately half to 7.9 mM on addition of 1 M NaCl. In contrast, the cmc of sodium n-dodecylsulphate decreases by approximately an order of magnitude (8 mM to 0.83 mM) on addition of 0.4 M NaCl due to the common ion effect.

As the data in Table 3.3 show, the decrease in cmc with acyl chain length is large, and if we extrapolate these cmcs to a chain length typical of a membrane lecithin such as dipalmitoylphosphatidylcholine (diC_{16} lecithin), we obtain a cmc in the region of 2×10^{-11} M. This suggests that, for all practical purposes, the concentration of monomeric phospholipids in biological systems is vanishingly small. Attempts have been made to directly measure the cmc of diC_{16} lecithin by filtration (Smith and Tanford, 1972). In this method, radioactively labeled phospholipid in aqueous methanolic solutions is rapidly filtered under pressure through cellulose filters. If the phospholipid is below its cmc, then the concentration in the filtrate will equal that in the original solution; if the phospholipid is above its cmc, the micelles will be retained by the filter and the concentration of phospholipid in the filtrate will fall. The cmc can be determined by filtering solutions covering a range of concentration and finding the concentration at which the filtrate is just depleted in phospholipid. By carrying out the experiment for solutions containing decreasing methanol concentrations, the cmcs can be extrapolated to zero methanol concentration to give an estimate of the cmc in pure water.

A value of 4.6×10^{-10} M was obtained, which is somewhat larger than the estimate based on extrapolation of the short-chain homologues. However, there is no doubt that for the majority of naturally occurring phospholipids and other membrane lipids, there is no appreciable concentration of monomeric species in biological systems. For this reason, the intracellular transfer of lipid molecules *in vivo* depends upon mediation by specific transfer proteins or vesicles, since the monomer concentrations are too low to account for "trafficking" of lipids as monomers.

3.1.4. Micellar-like Behavior of Some Protein Associations

As well as micelle formation by small amphipathic molecules, there are a number of protein associations that have been described in terms of micellization. Association or aggregation of proteins is a very common occurrence but cannot generally be described in terms of micellization with a well-defined cmc. More frequently the process occurs with a low cooperativity, the degree of association increasing with concentration, so that as the concentration of the solution is raised, dimers, trimers, tetramers, etc., form with increasing frequency. Such indefinite self-association is not described by a monomer-micelle equilibrium. However, for some proteins such as the caseins and tubulin, they associate cooperatively and the process can be described as a micellization with the formation of micelles of defined structure.

Caseins are major constituents of milk that contains approximately 4.2 wt% of protein. The other proteins in milk besides the caseins include α and β-lactoglobulin and various γ-globulins. The caseins in milk are present at room and body temperatures as micelles (100–300 nm in diameter) complexed with calcium salts (phosphates and citrates), the other particulate species being fat globules (triglycerides coated with protein and phospholipids) and lipoprotein particles. The casein protein is a mixture of several caseins that all have fairly similar "monomeric" molecular weights, specifically α_{s1}-casein (mol. wt. 2.36×10^4), β casein (mol. wt. 2.40×10^4), α_{s2}-casein (mol. wt. 2.52×10^4), and κ-casein (mol. wt. 1.90×10^4) which occur in casein micelles in the approximate ratio 4:4:1:1. There is also a small proportion of a proteolytic breakdown product of β-casein called γ-casein (Dickinson and Stainsby, 1982).

The origin of the formation of casein micelles lies in the primary sequence of the caseins. Thus, for example, β-casein has a primary sequence of 209 amino acid residues, the first 50 of which contain numerous acidic groups (aspartate and glutamate) and 5 phosphorylated serines, while residues 51–209 are largely hydrophobic. This amphipathic structure resembles that of a typical surfactant but of course on a much larger scale, as a sequence β-casein exhibits a cmc of approximately 4×10^{-2} wt% or 17 μM at pH 7, ionic strength 0.2 (Schmidt and Payens, 1976). Another structural feature of the caseins is the abundance of proline residues (~ 9 wt%), which inhibits the formation of secondary (helical) structure in the casein monomers so that they approximate more to random chains than many globular proteins of comparable molecular weight.

The equilibrium between monomeric and micellar β-casein is very markedly dependent on temperature and can be followed by ultracentrifugation. Figure 3.5a shows the sedimentation (schlieren) pattern for β-casein in which the peaks corresponding to monomeric and micellar β-casein can be clearly observed. As the temperature is lowered, the proportion of β-casein in micellar form decreases to less than 10 mole% at 10°C (Fig. 3.5b) (Evans et al., 1979). This behavior is characteristic of a hydrophobic association. Further support for the important role of the hydrophobic interactions in this system comes from chemical modification of β-casein and studies in deuterium oxide (^2H_2O). If β-casein

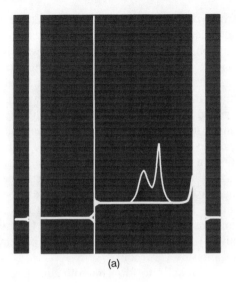

(a)

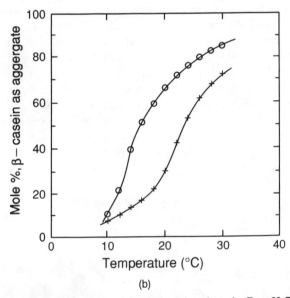

(b)

Fig. 3.5. a. Sedimentation pattern of β-casein (phosphate buffer pH 7, concentration 0.01 g cm^{-3}) at 20°C. The centrifugal field is from left to right. The right-hand larger peak corresponds to β-casein micelles and the left-hand peak to monomeric β-casein. **b.** The effect of temperature on the monomer-micelle equilibrium for β-casein derived from sedimentation data in **a.** $\bigcirc$, 0.01 g cm^{-3}; $\times$, 0.005 of cm^{-3} β-casein. Adapted from Evans et al. (1979).

is treated with carboxypeptidase A, three residues at the C-terminal (hydrophobic) end of the molecule are removed, i.e., residues 207–209 (Ile, Ile, Val). Reducing the hydrophobicity of the molecule in this way displaces the equilibrium toward monomer; thus at 20°C (concentration 0.01 g cm^{-3}), 67% of β-casein is micellar, but for the carboxypeptidase A–treated β-casein only 30% is micellar. In contrast, the equilibrium is displaced toward the micellar β-casein if the solvent is changed from H_2O to ^2H_2O. Although substitution of ^2H_2O for H_2O produces changes in all solvent interactions, the changes are particularly marked for hydrophobic hydration effects, so that the hydrophobic interaction is stronger in ^2H_2O than in water.

A further property of the caseins that is important in the context of milk is their ability to bind Ca^{2+} ions. In fact, the caseins can be subdivided on the basis of their solubility in Ca^{2+} solutions (Farrell et al., 1988). The α_s and β-casein are calcium-sensitive and can be salted out by Ca^{2+} ions, while κ-casein is calcium-insensitive. In milk the calcium-sensitive protein protects the other caseins from aggregation by Ca^{2+} ions.

The self-assembly of tubulin to form microtubules is a cooperative process having features that resemble micellization, with the resulting "micelle" having a precisely defined structure. The microtubules formed in self-assembly of tubulin are hollow cylinders of 25 nm outside diameter and 15 nm internal diameter. The walls of the cylinder are made up of the tubulin protein, which has two almost identical subunits, α-tubulin and β-tubulin, each of molecular weight 55,000. Microtubules are widely distributed *in vivo* in the cytoplasm of cells and are major components of eukaryotic cilia and flagella. *In vitro* tubulin will self-assemble only in the absence of Ca^{2+} to form either microtubules—in which the tubulin molecules are arranged on a helical surface lattice with 13 tubulin molecules per turn (Fig. 3.6)—or closed cyclic oligomers of molecular weight 3×10^6. The self-assembly can be monitored by tubidity measurements at 350 nm and by electron-microscopy and occurs in buffers containing Mg^{2+}, guanosine triphosphate (GTP), and glycerol, which stabilizes the microtubules. A typical buffer system for the formation of microtubules *in vitro* thus might consist of 0.1 M phosphate, 20 mM Mg^{2+}, 0.1 mM GTP,

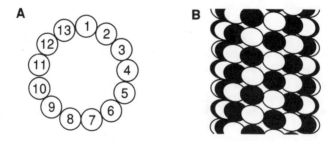

Fig. 3.6. Self-assembly of tubulin to form microtubules. **A.** Cross-section of the surface lattice of 13 tubulin subunits. **B.** Surface lattice of α- and β-tubulins (black and white ellipsoids). From Timasheff (1978), with permission of the publisher.

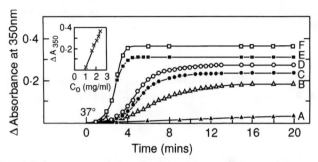

Fig. 3.7. Determination of the critical concentration of self-assembly of microtubules from tubulin at 37°C. The curves A–F correspond to increasing concentrations (see inset). From Timasheff (1978), with permission of the publisher.

1 mM ethylenebis (oxyethylenitrilo) tetraacetic acid (EGTA), and 3.6 M glycerol. The function of the EGTA is to complex Ca^{2+} ions which inhibit microtubule formation. On heating tubulin in such a buffer system to 37°C, the tubidity at 350 nm increases rapidly, and in a few minutes microtubules assemble with external diameters of around 30 nm, which is very similar to those found *in vitro* (Timascheff, 1978). The process is reversible; on cooling to 20°C, the microtubules rapidly dissociate to the tubulin subunits. If the change in absorbance at 37°C is plotted as a function of the tubulin concentration, it does not extrapolate to the origin but to a finite protein concentration (C_r) that is analogous to a cmc (Fig. 3.7). The value of C_r is dependent on the glycerol concentration: in the absence of glycerol, C_r at 37°C is 8 mg/ml; in 30% v/v glycerol, C_r decreases to 0.75 mg/ml. Tubulin association, like that of β-casein, is inhibited by low temperatures characteristic of hydrophobic interactions. The glycerol effect is believed to arise from a preferential exclusion of glycerol from the solvation shells of the tubulin monomers; thus the decreasing amount of water of hydration available per monomer with increasing glycerol concentration favors the self-assembly process, since in the microtubule a proportion of the tubulin-water contacts are replaced by hydrophobic tubulin-tubulin contacts.

3.2. THE SIZE AND SHAPE OF MICELLES

3.2.1. General Considerations

The size of micelles relates to the number of monomers per micelle or what is commonly called the aggregation number (N), which in the simplest case of the closed association model (see below) of micelle formation can be written

$$NS \rightleftharpoons S_N \qquad (3.9)$$

The molecular weight (M) of a micelle has been empirically related to the surfactant chain length (n) as follows (Barry and Russell, 1972)

$$\log M = d + en \tag{3.10}$$

where d and e are constants. Combining equation 3.10 with equations 3.3 and 3.6 leads to a log-log relationship between M and c_i.

$$\log M = f \log c_i + g \tag{3.11}$$

where f and g are constants. It follows that increasing the counterion concentration not only decreases the cmc (equation 3.6) but also increases the micellar molecular weight and aggregation number of ionic surfactants. The data in Table 3.1 for SDS show N increasing with salt concentration. For nonionic surfactants, it is also found that salts which depress the cmc also increase the micellar molecular weights in accordance with the lytotropic series, with the exception of salts with hydrophobic ions such as the tetramethylammonium salts.

Another important factor in relation to the molecular weight of micelles is the degree of polydispersity. In general, at low concentration just above the cmc smaller spherical micelles have a narrow distribution of molecular weight and approximate to a monodisperse distribution, but increasing the salt concentration and depressing the cmc, so that larger micelles are formed, leads to more polydisperse distributions. Furthermore, increasing the total surfactant concentration above the cmc generally leads to larger and more polydisperse micelles. Here the question of micellar shape is an important consideration. It is well established that in the vicinity of the cmc at low salt concentration anionic, cationic, and nonionic surfactant micelles mostly approximate to spheres, but on addition of salt and/or increasing the total surfactant concentration the micelles become asymmetric. Ignoring packing constraints, the sphere is the smallest shape with a given surface area per molecule and will be the thermodynamically favored shape. However, for micelles formed from a surfactant with a given hydrophobic chain length, there will be a critical length (l_c) that will limit the radius of the spherical micelle (R). For a spherical micelle of radius R and volume $V = 4/3\pi R^3$ and surface area $A = 4\pi R^2$, the volume (v) and area (a) *per amphiphile* are related by

$$\frac{v}{aR} = \frac{1}{3} \tag{3.12}$$

It is reasonable to assume that the mean area per amphiphile will always be close to an optimum value a_o, and since R cannot exceed l_c it follows that the critical condition for the formation of a spherical micelle is that

$$\frac{v}{a_o l_c} = \frac{1}{3} \tag{3.13}$$

and hence if $v/a_o l_c$ exceeds $\frac{1}{3}$ the micelle must adopt an alternative shape. The criteria for such alternative shapes are that no point within the structure can be further from the hydrophobic core–aqueous interface than l_c and that the micellar aggregation number $N = V/v = A/a_o$ (Israelachvili et al., 1976). A cylindrical micelle obeys such criteria. For a cylinder it is easily shown that

$$\frac{v}{a_o l_c} = \frac{1}{2} \tag{3.14}$$

and theoretically there is no limit on the length of the cylinder so that the aggregation number could go to infinity. In practice, cylindrical micelles occur with finite length; hence other factors must control their length. Other shapes could, however, exist for values of $v/a_o l_c$ between $\frac{1}{3}$ for spheres and $\frac{1}{2}$ for cylinders. Israelachvili et al. (1976) show that globules (essentially two fused spheres) of different degrees of asymmetry obey the above criteria with values of $v/a_o l_c > \frac{1}{3}$. A toroid is another possible shape which has a finite aggregation number, depending on its overall diameter.

3.2.2. Bile Salt Micelles

The determination of the aggregation numbers of bile salt micelles is a typical example of the problems associated with the characterization of ionic micelles in general. There are several physical techniques that can be used to characterize micelles in terms of molecular mass and size. They may be broadly classified into thermodynamic (equilibrium) methods such as membrane osmometry, static (elastic) light scattering, and sedimentation equilibrium and transport (hydrodynamic) methods such as sedimentation rate, viscosity, and diffusion methods. Diffusion coefficients can be determined by dynamic (quasi-elastic) light scattering. The thermodynamic methods are used to obtain molecular masses (M) (commonly called molecular weights) and hence the aggregation numbers from the known monomer molecular weights (M_o), i.e., $N = M/M_o$, whereas the transport methods yield the size of the micelles, often expressed as an equivalent hydrodynamic radius or Stokes radius (R_h) on the assumption that the micelle is spherical. The molecular weight relates to the unhydrated micelle, because the physical parameter (e.g., light intensity or osmotic pressure) is measured for solutions of known concentration made up from the *dry* solute. In contrast, the transport measurements relate to the movement of the micelle in solution, which is affected by the extent of hydration and asymmetry of the micelle. A further general point relating to micellar molecular weight is that the type of average molecular weight obtained depends on the particular physical technique used. The most commonly encountered averages are the number average ($\overline{M}_n$) defined by

$$\overline{M}_n = \frac{\Sigma M_i n_i}{\Sigma n_i} \tag{3.15}$$

where M_i and n_i refer to the molecular weight and number of species i in the system, and the weight average ($\overline{M}_w$), defined by

$$\overline{M}_w = \frac{\Sigma M_i w_i}{\Sigma w_i} \tag{3.16}$$

where w_i refers to the weight of species i in the system. For a micellar solution, M_i, n_i, and w_i refer to the molecular weight, number, and weight respectively of a micelle. For a monodisperse distribution of micelles, where only one size of micelle is present in solution, so that in principle its molecular weight will be the same however it is measured, $\overline{M}_w = \overline{M}_n$, but for a polydisperse distribution, $\overline{M}_w$ will exceed $\overline{M}_n$ and the ratio $\overline{M}_w/\overline{M}_n$ is a convenient measure of polydispersity. Of the thermodynamic techniques, osmometry yields $\overline{M}_n$ whereas static light scattering and sedimentation equilibrium yield $\overline{M}_w$.

Static light scattering is a very commonly used technique for the determination of molecular weights ($\overline{M}_w$) of micelles, but as with all the thermodynamic methods it is beset by the complexities of charged micellar systems in aqueous media. It is beyond the scope of this text to present light scattering theory; it is sufficient to state that the equation for the determination of molecular weight of a small particle from the measurement of the intensity of light scattered at 90° to an incident beam from its solution of concentration c_2 is given by

$$\frac{Kc_2}{R_{90}} = \frac{1}{M} (1 + 2A_2 c_2 \text{---}) \tag{3.17}$$

where R_{90} is the Rayleigh factor defined by $i_{90} r^2 / I_o$ where i_{90} and I_o are the intensity of light scattered at 90° and the incident beam intensity, respectively, and r is the distance from the scattering particles to the point of observation. The constant K is defined by

$$K = \frac{2\pi^2 n^2}{N\lambda^4} \left(\frac{\partial n}{\partial c_2}\right)^2_{m_3} \tag{3.18}$$

where n is the refractive index of the solution, N Avogadro's Constant, λ the wavelength of the light in vacuum and $(\partial n/\partial c_2)_{m_3}$ the refractive index increment of the solute of concentration c_2 at constant salt concentration (m_3). The parameter A_2 is the second virial coefficient and relates to the interactions between pairs of solute particles in solution; thus A_2 directly reflects the nonideality of the solution. For an ideal solution, $A_2 = 0$.

In principle, for uncharged particles of constant molecular weight (M), it is a relatively simple matter to determine Kc_2/R_{90} as a function of c_2 and obtain a linear plot that, when extrapolated to zero concentration ($c_2 \rightarrow 0$), gives an intercept 1/M and a slope $2A_2$ and hence the molecular weight of the solute and the second virial coefficient.

For a charged micellar system, a number of problems arise that complicate the application of this technique. First, the micelles are in equilibrium with monomers and only exist above the cmc; second, for many micellar systems the micelles increase in size as the total amphiphile concentration is increased— i.e., M depends on c_2; and third, there are preferential interactions between the micelles and solvent components (water and any salts present in the solution). The plot of R_{90} vs. c_2 shows a steep increase at the cmc, since the micellar solution scatters considerably more light than the monomer solution. Thus it is usual to replace c_2 by c_2-cmc and make all scattering measurements relative to the monomer solution at the cmc. Rigorously for an equilibrium (equation 3.9) the monomer concentration increases slightly above the cmc, but this is usually neglected (see sec. 3.2.3). The second problem is more difficult, because it is impossible to separate out the nonideality effects arising from micelle-micelle interactions when the micelles are increasing their size with increasing concentration. The third problem of preferential interactions should in principle be solved by dialyzing the micelles against solvent and replacing the refractive index increment $(\partial n/\partial c_2)_{m_3}$ with the increment $(\partial n/\partial c_2)_\mu$ at constant solvent chemical potential μ, but in practice this cannot be done for micellar solutions because dialysis disturbs the micelle-monomer equilibrium. To overcome this problem, Vrij and Overbeek (1962) showed that the apparent value of the molecular weight of a micelle M* can be related to the true value, M, by the relation

$$\sqrt{M^*} = \sqrt{M} + \frac{m_3(\partial n/\partial c_3)}{\sqrt{M}(\partial n/\partial c_2)_{m_3}}(1 - c_3\bar{v}_3)\left(\frac{\partial m_3}{\partial m_2}\right)_\mu \qquad (3.19)$$

where $(\partial n/\partial c_3)$ is the refractive index increment for the salt present of partial specific volume $\bar{v}_3$ and $(\partial m_3/\partial m_2)_\mu$ is the change in molality of salt with micellar solute at constant chemical potential. By use of equation 3.17 with c_2 replaced by c_2-cmc, an apparent micellar molecular weight, M*, can be obtained at the cmc for anionic micelles in a range of electrolytes having a common counterion at the same concentration, e.g., NaF, NaCl, NaBr, etc. Independent measurements of $(\partial n/\partial c_3)$ enable a plot of M* vs. $m_3(\partial n/\partial c_3)$ to be made to give an intercept $\sqrt{M}$, so giving the true molecular weight of the micelles. The extent of the preferential interactions between the electrolyte ions and the micelles is reflected in the slope of such a plot, from which $(\partial m_3/\partial m_2)_\mu$ can be obtained if $(1 - c_3\bar{v}_3)/(\partial n/\partial c_2)_{m_3}$ is measured in separate experiments. The primary assumption in such a treatment is that the co-ions—e.g., F^-, Cl^-, Br^-, etc.— have no effect on the molecular weight of the micelles.

The above discussion illustrates the problems required to be overcome in order to obtain reliable estimates of the minimum molecular weights of charged micelles at the cmc. These problems are not restricted to the light scattering technique but arise in a similar way in all the thermodynamic methods. The aggregation numbers of micelles quoted in the literature have not always been obtained by the above method, since numerous less rigorous methods which

TABLE 3.4. Aggregation Numbers of Sodium Deoxycholate Micelles as a Function of Na$^+$ Ion Concentration[a]

[Na$^+$] (M)	N $\pm$ (Standard Deviation)	No. of Studies
0	11 $\pm$ 8	7
0.01	6 $\pm$ 5	4
0.05	10 $\pm$ 8	3
0.10	13 $\pm$ 4	7
0.15	16 $\pm$ 3	9
0.20	17 $\pm$ 3	7
0.25	20 $\pm$ 4	4
0.30	22 $\pm$ 10	2
0.50	41 $\pm$ 16	4
0.60	69 $\pm$ 27	3

From Kratohvil (1986), with permission of the publisher.
[a]The data mostly relate to measurement at 20–25°C.

do not fully take into account increasing micellar size with concentration and preferential interactions have been used. The bile salts are no exception, and a comprehensive review of the aggregation numbers of sodium deoxycholate—the most studied bile salt—reveals very wide discrepancies between different reported results (Kratohvil, 1986). Table 3.4 shows the average aggregation numbers and standard deviations from a range of reported studies. It is clear that the micelles formed at low salt concentration are very small, but there is a trend toward much larger micelles with increasing salt concentration. The small size of the bile salt micelles is consistent with their structural features; one side of the molecule is hydrophobic and the other hydrophilic.

The variations of the apparent aggregation number N* (uncorrected for non-ideality effects) for the conjugated bile acids sodium taurocholate (NaTC) and taurodeoxycholate (NaTDC) with total concentration are shown in Figure 3.8. The scattering parameter Kc_2/R_{90} shows a relatively sharp transition at the cmc of NaTDC, and the transition occurs at a lower concentration in 0.6 M NaCl than in 0.15 M NaCl. For NaTC the cmc is less well defined and small aggregates are formed at the lowest concentrations. The apparent decrease in aggregation number at the highest concentration is due to positive second and higher virial coefficients due to intermicellar interactions. At low concentrations the negative curvature of the plots is due to concentration-dependent micellar growth.

3.2.3. Phospholipid Micelles

The short-chain phospholipids are sufficiently hydrophilic to form micelles in aqueous media in contrast to the long-chain membrane phospholipids, which form stable bilayers. The problems associated with the measurement of the size of bile salt micelles are also found with short-chain phospholipids. It is difficult,

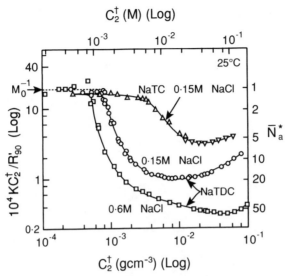

Fig. 3.8. Static light scattering plots Kc_2/M vs. c_2 and apparent aggregation numbers $\overline{N}_a^*$ for sodium taurocholate (NaTc) and taurodeoxycholate in aqueous solutions at 25°C. From Kratohvil 1986, with permission of the publisher.

if not almost impossible, to unambiguously disentangle the nonideality effects arising from micellar interactions in systems in which the micelles grow in size with increasing solute concentration to obtain the molecular weights and hence aggregation numbers unambiguously. Table 3.5 summarizes molecular weight data obtained from light scattering measurements on a homologous series of lecithins. The diC_6 lecithin forms small monodisperse micelles that change little with increasing salt concentration, whereas the diC_7 lecithin forms larger, more polydisperse micelles that grow considerably as the concentration of the lecithin is increased. The diC_8 lecithin exhibits phase separation in water and salt solutions—specifically sodium chloride and lithium iodide—so that at low temperatures (around room temperature) two aqueous phases are present, but on heating a single phase is found and at a critical lecithin concentration depending on the salt concentration, an upper consolute temperature occurs. This behavior is shown in Figure 3.9a and b. At low electrolyte concentrations (below 0.2 M), both sodium chloride and lithium iodide cause the upper consolute temperature to decrease. At higher salt concentrations the two salts behave differently, sodium chloride raising the upper consolute temperature and lithium iodide continuing to reduce it so that in 0.2 M of lithium iodide a single phase is formed on which it is possible to make light scattering measurements. The micelles in 0.2 M of lithium iodide are found to be very large and polydisperse and grow with increasing lecithin concentration. It is not clear what the shape of these micelles is; it is possible that they are extended rods, although they may form an extended branched network (Tausk, 1974c).

TABLE 3.5. Micelle Molecular Weights and Aggregation Numbers of Short-Chain Lecithins in Aqueous Media at 25°C

Compound		$\overline{M}_w$	N_w^a	$\overline{M}_w/\overline{M}_n$
	[NaCl]M			
Dihexanoylecithin (diC$_6$)	0	15400^b	33	1.06
Dihexanoylecithin (diC$_6$)	1	16200^b	34	—
Dihexanoylecithin (diC$_6$)	2	16350^b	35	—
Dihexanoylecithin (diC$_6$)	3	16300^b	34	1.1
	[Lecithin]^c mg/ml			
Diheptanoyllecithin (diC$_7$)	0.7	20,000	40	1.1
Diheptanoyllecithin (diC$_7$)	4	52,000	105	1.7
Diheptanoyllecithin (diC$_7$)	8	66,000	130	1.7
Diheptanoyllecithin (diC$_7$)	12	74,000	150	1.5
Diheptanoyllecithin (diC$_7$)	16	77,000	155	1.5
Dioctanoylecithin (dC$_8$)	0.128^d	0.32 × 10^6	605	Polydisperse
	10	2.38 × 10^6	4500	Polydisperse

Data from Tausk et al. (1974b,c).
[a]Weight-average aggregation number.
[b]M_w at the cmc.
[c]Solvent 10 mM phosphate, pH 6.9.
[d]Solvent 0.2 M lithium iodide.

3.2.4. β-Casein Micelles

The sizes of β-casein have been studied using several physical methods, including sedimentation rate, dynamic light scattering, and small-angle X-ray scattering and electron microscopy (Evans et al., 1979; Kajiwara et al., 1988). The β-casein association can be expressed by the equilibrium

$$i\beta \rightleftharpoons \beta_i \tag{3.20}$$

The most dramatic feature of this association is its temperature dependence. The value of i increases rapidly with temperature, although i also depends on the ionic strength, pH, and β-casein concentration.

In the temperature interval 10–40°C in phosphate buffer pH 7 (ionic strength ~ 0.08), sedimentation rate measurements in combination with intrinsic viscosity data gave the following dependence of aggregation number (n) on temperature (t°C)

$$n = 0.6t + 2 \tag{3.21}$$

from which it follows that the aggregation number increases from 14 to 26 between 20°C and 40°C for a 1 wt% solution (Evans et al., 1979). At higher ionic strength (0.2 M phosphate, pH 6.7), the aggregation number at 20°C

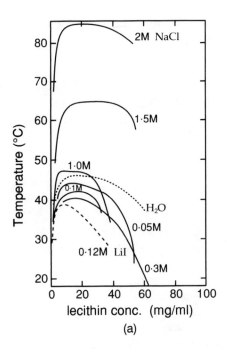

(a)

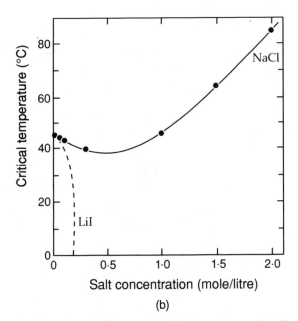

(b)

Fig. 3.9. a. Phase separation of dioctanoylecithin in aqueous salt solution. **b.** Effect of salt concentration on the upper consolute temperature. From Tausk et al. (1974), with permission of the publisher.

increases to 49, corresponding to a micellar molecular weight of approximately 1.2×10^6. Electron microscopy showed that the micelles were spherical with a diameter of approximately 340 Å. Small-angle X-ray scattering from β-casein micelles in 0.2 M phosphate (pH 6.7) is very well interpreted in terms of an oblate ellipsoid. For such a triaxial body the dimensions are defined in terms of three semi-axes a, b, and c. Table 3.6 gives the values of the semi-axes as a function of β-casein concentration at 25°C, together with the radius of gyration that for an ellipsoid is given by

$$R_g = \left(\frac{a^2 + b^2 + c^2}{5} \right)^{1/2} \tag{3.22}$$

Note that the square of the radius of gyration is the mass average of the square of the distance (r_i) of each mass element (m_i) from the center of mass of the body, i.e., $R_G^2 = \Sigma m_i r_i^2 / \Sigma m_i$.

The data in Table 3.6 show that β-casein micelles do not change their dimensions greatly with concentration in the range of 5–50 mg ml^{-1}. The increase in size with temperature is primarily due to expansion of the c-axis; for example, at a concentration of 5 mg ml^{-1} the c-axis increases by 179% between 5°C and 25°C, whereas the corresponding increases in the a and b axes are only 5.8% and 5.5%, respectively.

The hydrodynamic radii (R_h) of β-casein micelles as measured by dynamic light scattering are considerably larger than the radii of gyration; at 25°C R_h = 16.3 nm. A possible explanation of the difference between R_h and R_G is that the highly charged N-terminal region (the first 50 amino acid residues) forms a soft layer covering the surface of the ellipsoid, which has a rigid core composed of the hydrophobic C-terminal residues. The soft layer contributes to the hydrodynamic behavior of the micelles while making little contribution to the small-angle X-ray scattering and hence to R_G.

TABLE 3.6. Dimensions of β-Casein Ellipsoidal Micelles in 0.2 M Phosphate Buffer pH 6.7 at 25°C

β-Casein Concentration (mg/ml)	Semi-axes (nm)			R_G (nm)
	a	b	c	
5	11.43	11.93	5.32	7.75
10	12.03	11.85	5.55	7.95
20	12.83	12.63	5.65	8.44
30	13.08	12.88	5.82	8.61
40	12.85	12.66	5.93	8.49
50	12.33	12.66	5.95	8.34

Adapted from Kajiwara et al. (1988), with permission of the publisher.

3.3. THERMODYNAMICS OF MICELLIZATION

As discussed above, micelles form with a range of sizes and size distribution (polydispersity). In the case of micelles formed from ionically charged species, the sizes of the micelles and to some degree their polydispersity are influenced by the presence of electrolytes. Temperature also has an important effect on the position of the equilibria between monomeric and micellar species. The first problem to be addressed in considering the thermodynamics of micelle formation is how best to represent the equilibria between monomer and micelles. There are two basic models that can be used. The first is the closed association model, which has already been used above (equations 3.9), and for an unchanged monomer (S) is represented by the equation

$$NS \rightleftharpoons S_N \tag{3.23}$$

This implies that only one type of micelle consisting of a fixed number of monomers is formed. This is clearly an approximation but can nevertheless be a useful initial approach, provided the aggregation number (N) is large. Secondly and more rigorously, micellization can be represented by a multiple equilibrium model according to a range of equilibria, which can be represented as

$$S + S \overset{K_2}{\rightleftharpoons} S_2$$
$$S_2 + S \overset{K_3}{\rightleftharpoons} S_3$$

etc.

so that

$$S_{N-1} + S \overset{K_N}{\rightleftharpoons} S_N \tag{3.24}$$

There are two particular cases based on equation 3.2.4: case 1, when all the K^s are equal—i.e., $K_2 = K_3 = \ldots K_N$—which will give rise to a very wide distribution of micelle sizes; and case 2, when K_N is strongly dependent on N, so that one particular value of K_N is larger than all others. In this case, we have a distribution of K_N^s passing through a maximum for a particular K_N, although, unlike the closed association model, K_{N-1} and K_{N+1} etc. have significant (nonzero) values.

The standard Gibbs energy changes for the formation of a micelle of aggregation number N can be simply derived for the closed association model as follows. Using c_m and c_s to represent the concentrations of micelles and monomer respectively for equation 3.23, the equilibrium constant K is given by

$$K = \frac{c_m}{c_s^N} \tag{3.25}$$

$$\Delta G^{\circ} = -RT \ln K \qquad (3.26)$$

$$= -RT \ln c_m + NRT \ln c_s \qquad (3.27)$$

Hence at the cmc, c_s = cmc and the standard Gibbs energy change per mole of monomer (ΔG°) is given by

$$\Delta G^{\circ}_m = \frac{\Delta G^{\circ}}{N} = -\frac{RT}{N} \ln c_m + RT \ln cmc \qquad (3.28)$$

If N is large, the second term on the right-hand side will dominate; hence

$$\Delta G_m = RT \ln cmc \qquad (3.29)$$

The equilibrium constant can be written in terms of the total amphiphile concentration (c_T) as

$$K = \left(\frac{c_m}{c_T - Nc_m} \right) \qquad (3.30)$$

from which it follows that

$$\frac{dc_m}{dc_T} = \frac{K^{1/N}}{\left[NK^{1/N} + \frac{1}{N} c_m^{(1-N)/N} \right]} \qquad (3.31)$$

Figure 3.10 shows plots of dc_m/dc_T as a function of c_T for different values of N that demonstrate the increasing sharpness of micelle formation with increasing N. As N $\rightarrow$ ∞ the process is analogous to a phase separation.

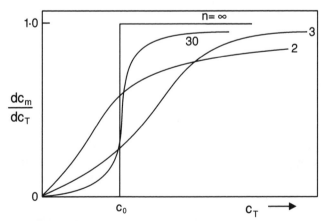

Fig. 3.10. Dependence of dc_m/dc_T on c_T and micelle aggregation number (N). From Hunter (1978), with permission of Oxford University Press.

For the multiple equilibrium model it follows from equation 3.24 that

$$K_N = \frac{c_m^{(N)}}{K_2 K_3 \dots K_{N-1} c_s^N} \qquad (3.32)$$

where $c_m^{(N)}$ is the concentration of micelles of aggregation number N, or

$$K_N^* = \prod_{N=2}^{N-1} K_N = \frac{c_m^{(N)}}{c_s^N} \qquad (3.33)$$

and hence

$$\Delta G_m^o = -\frac{RT}{N} \ln K_N^* \qquad (3.34)$$

This expression leads to equation 3.29, as for the closed association model when c_s = cmc. The problem of using the multiple equilibrium model is that of obtaining values of K_N as a function N; in general this information is not available and hence, although the model is more rigorous in practical terms, the Gibbs energy of micellization is usually calculated from the cmc using equation 3.29 and is assumed to apply to the formation of micelles of average aggregation number N.

For charged amphiphiles, counterion binding must be taken into account. This can be done by including the counterion in the equilibrium. Thus for an anionic amphiphile (S^-) with counterion (M^+)

$$NS^- + NM^+ \rightleftharpoons S_N M_m^{(N-m)-} + (N - m)M^+ \qquad (3.35)$$

and hence

$$K = \frac{c_m c_{M+}^{(N-m)}}{c_s^N c_{M+}^N} \qquad (3.36)$$

giving

$$\Delta G^o = -RT\{\ln c_m + (N - m)\ln c_m + N \ln c_s - N \ln c_{M+}\} \qquad (3.37)$$

$$\Delta G_m = \frac{\Delta G^o}{N} = -RT\left\{\frac{1}{N}\ln c_m + \frac{N-m}{N}\ln c_{M+} - \ln c_s - \ln c_{M+}\right\}$$

$$\qquad (3.38)$$

For large micelles $(\ln c_m)/N$ is negligible, so that at the cmc $c_s = c_{M+}$ = cmc.

Further, $(N - m)/N = \alpha$, the degree of dissociation of the micelles, hence

$$\Delta G_m^o = -RT[\alpha \ln \text{cmc} - 2 \ln \text{cmc}] \tag{3.39}$$

$$\Delta G_m^o = (2 - \alpha)RT \ln \text{cmc} \tag{3.40}$$

In the presence of added electrolyte of concentration x, $c_{M^+} = \text{cmc} + x$, hence

$$\Delta G_m^o = -RT[\alpha \ln (\text{cmc} + x) - \ln \text{cmc} - \ln (\text{cmc} + x)] \tag{3.41}$$

giving

$$\Delta G_m^o = RT[\ln \text{cmc} + (1 - \alpha) \ln (\text{cmc} + x)] \tag{3.42}$$

The Gibbs energies of micelle formation can thus be calculated for nonionic amphiphiles (equation 3.29) and ionic amphiphiles in the absence (equation 3.40) or presence (equation 3.42) of added electrolyte. The remaining issue is to define the standard state. The chemical potential of monomeric amphiphile assuming ideality may be written

$$\mu_1 = \mu_1^o + RT \ln x_1 \tag{3.43}$$

where x_1 is the mole fraction of amphiphile and the chemical potential in the standard state (μ_i^o) thus relates to the pure amphiphile. The chemical potential of the micellar amphiphile per mole of monomer may be written

$$\mu_{\text{mic}} = \mu_{\text{mic}}^o + \frac{RT}{N} \ln \frac{x_N}{N} \tag{3.44}$$

where x_N is the mole fraction of micellar amphiphile, hence x_N/N is approximately the mole fraction of micelles of aggregation number N. Rigorously, if n_{mic} and n_{H_2O} are the moles of amphiphile in micellar form and water, respectively, the mole fraction of micelles would be

$$x_{\text{mic}} = \frac{n_{\text{mic}}/N}{n_{\text{mic}}/N + n_{H_2O}} = \frac{n_{\text{mic}}}{n_{\text{mic}} + N n_{H_2O}} \tag{3.45}$$

whereas the mole fraction of micellar amphiphile would be

$$x_N = \frac{n_{\text{mic}}}{n_{\text{mic}} + n_{H_2O}} \tag{3.46}$$

It follows from equations 3.45 and 3.46 that if $n_{\text{mic}} \ll n_{H_2O}$, which is usually the case, then $x_{\text{mic}} = x_N/N$. The standard state according to equation 3.43

corresponds to unit mole fraction of micelles. Thus the Gibbs energy of micellization follows from equations 3.43 and 3.44, since for the equilibrium between monomer and micelles, $\mu_1 = \mu_{mic}$; hence

$$\mu_1^o + RT \ln x_1 = \mu_{mic}^o + \frac{RT}{N} \ln \frac{x_N}{N} \tag{3.47}$$

and

$$\Delta G_m^o = \mu_{mic}^o - \mu_1^o = RT \ln x_1 + \frac{RT}{N} \ln \frac{(x_n)}{N} \tag{3.48}$$

The second term on the right-hand side of equation 3.48 is negligible at the cmc, hence

$$\Delta G_m^o = RT \ln x_1 = RT \ln cmc \tag{3.49}$$

The standard-state issue thus resolves into the concentration units used for the cmc in equation 3.49. The importance of this can be seen by considering the micellization of sodium n-dodecylsulphate in water at 25°C. The cmc is 8×10^{-3} M, which would give from equation 3.40 with $\alpha = 0.18$ a value of -21.78 kJ mole^{-1} at 25°C for ΔG_m^o; in terms of mole fraction, the cmc is $8 \times 10^{-3}/(8 \times 10^{-3} + 997.7/18.015) = 1.444 \times 10^{-4}$ giving $\Delta G_m^o = -39.90$ kJ mole^{-1}. The standard-state issue has been discussed numerous times (see Hunter, 1987). Hunter (1987) suggests that it is physically more reasonable to consider micellization from the monomeric hydrate (at unit mole fraction). This makes no physical difference to the calculation of ΔG_m^o from equations 3.29, 3.40, and 3.42, and it appears to eliminate the problem of the unrealistic concept of an anhydrous pure micelle as defined by equation 3.44.

The calculation of the standard enthalpy and entropy of micellization follows from the application of the Gibbs-Helmholtz equation to the appropriate equation for ΔG_m^o depending on the nature (i.e., nonionic/ionic) of the amphiphile and the solvent compositions (i.e., $\pm$ salt).

$$\left(\frac{\partial \Delta G_m^o / T}{\partial 1/T} \right)_P = \Delta H_m^o \tag{3.50}$$

For small temperature intervals, assuming the aggregation number of the micelles remains constant, for nonionic amphiphiles

$$\Delta H_m^o = R \left(\frac{\partial \ln cmc}{\partial 1/T} \right)_P = -RT^2 \left(\frac{\partial \ln cmc}{\partial T} \right)_P \tag{3.51}$$

and for ionic amphiphiles assuming the degree of ion binding $(1 - \alpha)$ remains constant, in the absence of salt

$$\Delta H_m^o = -(2 - \alpha)RT^2 \left(\frac{\partial \ln \text{cmc}}{\partial T}\right)_P \qquad (3.52)$$

The entropy of micellization can hence be obtained from

$$\Delta G_m^o = \Delta H_m^o - T\Delta S_m^o \qquad (3.53)$$

In principle, the temperature coefficient of the cmc is all that is required to obtain ΔH_m^o and ΔS_m^o; in practice, the assumption that neither N nor α change with temperature restricts the use of the above equations and the direct measurement of the enthalpy of micellization by calorimetry is preferable.

Table 3.7 shows the thermodynamic parameters ΔG_m^o, ΔH_m^o, and $T\Delta S_m^o$ for a selection of compounds. A very comprehensive listing of thermodynamic parameters for micellization has been given by Kresheck (1975). It is significant that despite the chemical differences between these compounds the Gibbs energies of micellization are not widely different and lie in a range from approximately -20 to -35 kJ mol^{-1}. Apart from β-casein, the enthalpies of micellization are small compared with $T\Delta S_m^o$. Kresheck (1975) showed that for a

TABLE 3.7. Thermodynamic Parameters for Micellization[a]

Compound	Solvent/Temperature	ΔG_m^o	$\Delta H_m^o(M)$[b] kJ mole^{-1}	$T\Delta S_m^o$
Sodium n-dodecylsulphate[c]	H$_2$O/25°C	−21.80	0.38(C)	22.18
n-Dodecyltrimethylammonium[d] bromide	H$_2$O/25°C	−35.59	−1.39(C)	34.20
n-Dodecyltrimethylammonium[d] bromide	0.0175M NaBr/25°C	−35.26	−1.58(C)	33.68
n-Dodecyltrimethylammonium[d] bromide	0.05M NaBr/25°C	−34.92	−1.85(C)	33.07
n-Dodecyltrimethylammonium[d] bromide	0.1M NaBr/25°C	−34.79	−2.05(C)	32.74
PEG(9-10)p-t-octylphenol (TritonX 100)[e]	H$_2$O/25°C	−30.58	+8.79(T)	39.37
Sodium taurocholate[f]	H$_2$O/20°C	−24.11	−1.30(T)	22.81
Sodium deoxytaurocholate[f]	H$_2$O/20°C	−25.64	−4.03(T)	21.61
Dihexanoyl lecithin (diC$_6$)[g]	H$_2$O/25°C	−20.42	+6.69(C)	27.11
β-casein[h]	Phosphate pH7/25°C	−31.81	65.80(C)	97.61

[a]The data relate to a standard state of unit mole fraction of monomer. This was not always used by the original authors, and in such cases the values of ΔG_m^o and $T\Delta S_m^o$ were recalculated.
[b]Method of determination: C = calorimetry; T = temperature coefficient of the cmc.
Data from: [c]Pilcher et al. (1969); [d]Espada et al. (1970); [e]Ray and Nemethy (1971); [f]Carey and Small (1969); [g]Johnson et al. (1981); [h]Evans et al. (1979).

TABLE 3.8. Thermodynamics of Micellization of *n*-Dodecyltrimethylammonium Bromide in Water at Various Temperatures

Temperature °C	cmc (M)[a]	cmc (mole fraction)	α^a	ΔG_m^o	ΔH_m^o	$T\Delta S_m^o$
					(kJ mole^{-1})[b]	
5	0.0155	2.792×10^{-4}	0.195	-34.16	$+5.29$	39.45
15	0.0153	2.758×10^{-4}	0.206	-35.22	$+4.46$	39.68
25	0.0147	2.655×10^{-4}	0.256	-35.59	-1.39	34.20
40	0.0163	2.959×10^{-4}	0.250	-37.02	-4.60	32.42
50	0.0187	3.416×10^{-4}	0.286	-37.33	-4.88	32.45

[a]Data from: Ingram and Jones (1969); [b]Espada et al. (1970).

wide variety of ionic and nonionic surfactants in water and other liquids, a plot of ΔH_m^o vs. $T\Delta S_m^o$ was linear. Such plots are characteristic of compensation between enthalpy and entropy. While $T\Delta S_m^o$ is always positive in water, ΔH_m^o is negative for small values of $T\Delta S_m^o$ and positive for larger ones. The large increase in entropy on micellization in water is characteristic of the hydrophobic interaction and is generally considered to arise as a consequence of the loss of structural water when the hydrophobic chains of the amphiphile pass from the aqueous phase to the micellar interior. The fact that the increase in $T\Delta S_m^o$ is accompanied by a proportional enthalpy change, so that the larger the $T\Delta S_m^o$ the larger (more positive) the ΔH_m^o reflects the fact that energy is adsorbed on breaking down water structure. The concept of water-structure breakdown on micellization as the major driving force for micelle formation has, however, been questioned. For some amphiphiles, micellization still occurs at high temperature in water ($>90°C$) where most of the structure of water has already been lost due to the high thermal energy. Micellization also occurs in solvents such as hydrazine, which is not known to form structural layers around hydrophobic groups (Evans et al., 1984).

It is significant that for many amphiphiles, the enthalpy of micellization changes sign in the region of 20–30°C, corresponding to a minimum in the plot of cmc vs. temperature (see equation 3.51). Table 3.8 shows a typical example of the variation ΔG_m^o, ΔH_m^o, and $T\Delta S_m^o$ with temperature. The cmc passes through a minimum at 25°C; below this temperature ΔH_m^o is endothermic and at higher temperatures the enthalpy is exothermic. $T\Delta S_m^o$ increases as ΔH_m^o becomes more positive; a compensation plot of ΔH_m^o vs. $T\Delta S_m^o$ is linear with a slope of 1.2.

3.4. MIXED AMPHIPHILE SYSTEMS

Both in many commercial applications and *in vivo* (e.g., bile salts), amphiphiles are not present as single components but are frequently present as amphiphile mixtures, resulting in the formation of mixed micelles at low concentrations and complex phase behavior at high concentrations where water is a limiting component. There are numerous models for describing the behavior of mixed surfactant systems (see, e.g., Kamrath and Franses, 1986). The simplest case is that of two nonionic surfactants that mix ideally in the micelle. For such a system Clint (1975) showed, using the phase separation model, that the cmc for the formation of the mixed micelle (C_m) is given by the expression

$$\frac{1}{C_m} = \sum_i \frac{x_i}{c_i} \tag{3.54}$$

where c_i is the cmc of component i at mole fraction x_i. The behavior of mixtures of nonionic, nonionic-ionic, and ionic-ionic surfactants when the surfactants do not mix ideally in the mixed micelle can become very complex, as discussed by Kamrath and Franses (1986). Mysels (1978) considered the case of two surfactants, A and B, A having the higher cmc and forming micelles in which B was freely soluble. However, A was only sparingly soluble in B micelles. Thus A and B deviate from ideal mixing in the micellar phase. The phase diagram for this situation is illustrated in Figure 3.11. Starting on the left of the diagram, the addition of B to A reduces the cmc at which micelles rich in A will form. At high mole fraction of B, the B-rich micelles form. However, near the middle of the phase diagram—e.g., $x_B \sim 0.6$—or increasing the total surfactant concentration after the formation of B-rich micelles, the concentration of monomeric A increases until it reaches its cmc, above which both A-rich

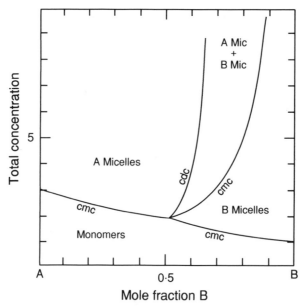

Fig. 3.11. Phase diagram for a mixture of surfactants A and B which do not rise ideally in the micellar phase. From Mysels (1978), with permission of the publisher.

and B-rich micelles coexist, each saturated with the other. As the total surfac-tant concentration is further increased, the A-rich micelles solubilize the B-rich micelles until a critical point (the critical demicellization concentration [cdc]) at which the B-rich micelles disappear, leaving only A-rich micelles. This type of phenomenon is predicted for mixtures of both nonionic and ionic micelles when the hydrophobic moieties have a limited mutual solubility, such as might be found for mixtures of hydrocarbons and perfluorocarbons.

In practical terms, the physiological function of the bile salts is dependent on the formation of mixed micelles with phospholipids. When food enters the small intestine, it triggers the release of the digestive hormones cholecystokinin and secretin. Cholecystokinin initiates the contraction of the gall bladder and secretion of bile into the small intestine and bile duct and the secretion of digestive enzymes, while secretin causes the pancreas to secrete sodium bicar-bonate. Food lipids, particularly lecithin and cholesterol, are solubilized by the bile salts in mixed micelles, enabling the pancreatic lipases to break down the phospholipid. The bile salts also play a regulatory role for a number of cho-lesterol-metabolizing enzymes in the liver and intestine. The products of pan-creatic hydrolysis—e.g., fatty acids in the mixed micelles—are transported across the unstirred layer of the intestinal mucosa surface and hence to the epithelial cells for adsorption. The bile salts themselves are conserved by a recycling mechanism through which they are absorbed in the small bowel and are returned via the portal drainage system to the liver (enterohepatic circula-tion).

In contrast to synthetic surfactants, the ternary systems of bile salts/lipid/water display larger zones of mixed micelle stability. The structure of these mixed micelles (Fig. 3.12) depends on the "sidedness" of the bile salt struc-

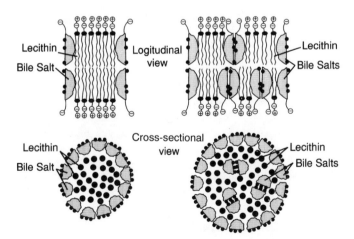

Fig. 3.12. Structure of mixed micelles of bile salts and phospholipid. Some of the bile salt may exist as dimers within the mixed micelle. From Mazer et al. (1980), with permission of the American Chemical Society.

ture, so that the hydrophobic side of the molecules interacts with lipid bilayer fragments. In this way a relatively small number of bile salt molecules can solubilize a large number of phospholipid molecules, while at the same time making the phospholipid headgroups accessible to the action of lipases.

3.5. SURFACTANTS IN THE ENVIRONMENT

The widespread use of surfactants in detergent formulation for herbicides, pesticides, and industrial and domestic cleaning agents inevitably leads to a degree of pollution of the environment by surfactants. The transition from the use of soaps to synthetic surfactants occurred in the late 1940s and early 1950s, when it became economically feasible to use alkylbenzene sulphonates (ABS). Initially ABS was derived from tetrapropylene and contained a proportion of branched chains that are not degraded by microorganisms; once this fact was established, ABS was replaced by linear alkylbenzene sulphonates, which are biodegradable and hence much less of a pollution hazard (Swisher, 1987). Although surfactants are in general nontoxic to humans, the World Health Organization (WHO) set a limit of 0.2 ppm of surfactant in drinking water in 1970, and this limit was also set for environmental waters by the United States in 1973. The concentrations of surfactants found in the environment are generally in the range of 20–60 μg per liter, although in untreated sewage values as high as 14 mg per liter have been found. Low levels of surfactants (μg per liter) will not in general endanger aquatic life; for fish the LD_{50} value (i.e., the concentration resulting in the death of 50% of a population in a period of 96 hours) is typically 10–20 mg per liter. Cationic surfactants are often more toxic than anionic or nonionic surfactants; however, they are used on a smaller scale. The site of attack by surfactants on fish is the gill. The gills of teleost fish are the sites of gas exchange and regulation of ionic and osmotic balance; they present a large surface area to the environment and hence are exposed to a wide range of potentially toxic environmental contaminants. The cells of the gill epithelia are only a few layers thick. There are several types of epithelial cells, including mucous secreting cells, that keep the epithelium covered with a protective layer of mucus. While surfactants can effect the permeability of gills to water, damage the structure of the gill surface, and rupture the epithelial cells, such effects occur at concentrations very much higher than those found in the environment (Partearroyo et al., 1991). There is an increasing awareness of environmental pollution, but relative to other forms of pollution surfactants do not in general present any great threat.

REFERENCES AND FURTHER READING

Adamson, A.W. (1960) Physical Chemistry of Surfaces. Interscience, New York.

Barry, B.W. and Russell, G.F.J. (1972) Prediction of micellar molecular weights and thermodynamics of micellization of mixtures of alkyltrimethylammonium salts. J. Coll. Int. Sci. *40*, 174–194.

Bijsterbosch, B.H. and Van den Hul, H.J. (1967) Comments on the paper "Gibbs equation for the adsorption of organic ions in presence and absence of neutral salt," by D.K. Chattoraj. J. Phys. Chem. *71*, 1169–1170.

Carey, M.C. and Small, D.M. (1969) Micellar Properties of dihydroxy and trihydroxy bile salts: Effects of counterion and temperature. J. Coll. Int. Sci. *31*, 382–396.

Clint, J. H. (1975) Micellization of mixed nonionic surface active agents. J. Chem. Soc. (Faraday Trans. 1) *71*, 1327–1334.

Dickinson, E. and Stainsby, G. (1982) Colloids in Food. Applied Science, London and New York, chap. 8.

Elworthy, P.H. and Mysels, K.J. (1966) The surface tension of sodium *n*-dodecylsulphate solutions and the phase separation model of micelle formation. J. Coll. Int. Sci. *21*, 331–347.

Espada, L., Jones, M.N. and Pilcher, G. (1970) Enthalpy of micellization. Part 2: *n*-Dodecyltrimethylammonium bromide. J. Chem. Thermodynam. *2*, 1–8.

Evans, M.T.A., Phillips, M.C. and Jones, M.N. (1979) The conformation and aggregation of bovine β-casein A. Part 2: Thermodynamics of thermal association and the effects of changes in polar and apolar interactions on micellization. Biopolymers *18*, 1123–1140.

Evans, D.F., Allen, M., Ninham, B.W. and Fonda, A. (1984) Critical micelle concentrations for alkyltrimethylammonium bromides in water from 25° to 160°C. J. Sol. Chem. *13*, 87–101.

Farrell, H.M., Kumosmski, T.F., Pulaski, P. and Thompson, M.P. (1988) Calcium-induced associations of the caseins: A thermodynamic linkage approach to preparation and resolubilization. Arch. Biochem. Biophys. *265*, 146–158.

Gibbs, J.W. (1931) The Collected Works of J.W. Gibbs. Vol. 1. Longemans, Green, New York, p. 301.

Hartley, G.S. (1936) Aqueous solutions of paraffin chain salts. Hermann, Paris.

Helenius, A. and Simons, K. (1975) Solubilization of membranes by detergents. Biochim. Biophys. Acta *415*, 29–79.

Hofman, A.F. and Mysels, K.J. (1988) Bile salts as biological surfactants. Colloids Surf. *30*, 145–173.

Hunter, R.J. (1987) Foundation of Colloid Science. Vol. 1. Clarendon Press, Oxford, chap. 10.

Ingram, T. and Jones, N.N. (1969) Membrane potential studies on surfactant solution. J. Chem. Soc. (Faraday Trans. 1) *65*, 297–304.

Israelachvili, J.N., Mitchell, D.J. and Ninham, B.W. (1976) Theory of self-assembly of hydrocarbon amphiphiles into micelles and bilayers. J. Chem. Soc. (Faraday Trans. 2) *72*, 1525–1568.

Johnson, R.E., Wells, M.A. and Rupley, J.A. (1981) Thermodynamics of dihexanoylphosphatidylcholine aggregation. Biochemistry *20*, 4239–4242.

Jones, M.N. (1975) Biological Interfaces. Elsevier, Amsterdam.

Jones, M.N. and Piercy, J. (1972) Light scattering studies on *n*-dodecyltrimethylammonium bromide and *n*-dodecylpyridinium iodide. J. Chem. Soc. (Faraday Trans. 1) *68*, 1839–1848.

Jones, M.N. and Reed, D.A. (1968) The effect of electrolytes on the critical micelle concentration of some cationic surfactants. Proc. Int. Vth Congr. Surface Activity, 1081–1089.

Kajiwara, K., Niki, R., Urakawa, H., Hiragi, Y., Donkai, N. and Nagura, M. (1988) Micellar structure of β-casein observed by small-angle X-ray scattering. Biochim. Biophys. Acta 955, 128–134.

Kamrath, R.F. and Franses, E.I. (1986) in Phenomena in Mixed Surfactant Systems. J.F. Scamehorn, Ed. ACS Symposium Series no. 311, Washington, chap. 3.

Kratohvil, J.P. (1980) Comments on some novel approaches for the determination of micellar aggregation numbers. J. Coll. Int. Sci. 75, 271–275.

Kratohvil, J.P. (1986) Size of the bile salt micelles: Techniques, problem and results. Adv. Coll. Int. Sci. 26, 131–154.

Kresheck, G.C. (1975) in Water, a Comprehensive Treatise. Vol. 4. F. Franks, Ed. Plenum Press, New York, pp. 95–167.

Mazer, N.A., Benedek, G.B. and Carey, M.C. (1980) Quasielastic light scattering studies on aqueous biliary lipid systems. Mixed micelle formation in bile salt-chain solutions. Biochemistry 19, 601–615.

Mukergee, P. and Mysels, K.J. (1971) Critical micelle concentrations of aqueous surfactant systems. Nat. Stand. Ref. Data. Ser., Nat. Bur. Stand. (U.S.) no. 36.

Mysels, E.K. and Mysels, K.J. (1965) Conductimetric determination of the critical micelle concentration of surfactants in salt solution. J. Coll. Sci. Int. 20, 315–321.

Mysels, E.K. and Mysels, K.J. (1972) Interpretation of the conductivity of micelles and other electrolytes dissolved in concentrated salt solution. J. Coll. Int. Sci. 38, 388–394.

Mysels, K.J. (1978) Critical demicellization concentration? J. Coll. Int. Sci. 66, 331–334.

Parfitt, G.D. and Smith, A.L. (1962) Conductivity of sodium dodecylsulfate solutions below the critical micelle concentration. J. Phys. Chem. 66, 942–943.

Partearroyo, M.A., Pilling, S.J., Hammond, K. and Jones, M.N. (1991) The lysis of isolated fish (*Oncorhychus mykiss*) gill epithelial cells of surfactants. Comp. Biochem. Physiol. 100C, 381–388.

Pilcher, G., Jones, M.N., Espada, L. and Skinner, H.A. (1969) Enthalpy of micellization. Part 1: Sodium *n*-dodecylsulphate. J. Chem. Thermodynam. 1, 381–392.

Ray, A. and Némethy, G. (1971) Micelle formation by nonionic detergents in water-ethylene glycol mixtures. J. Phys. Chem. 75, 809–815.

Roda, A., Hofmann, A.F. and Mysels, K.J. (1983) The influence of bile salt structure on self-associated in aqueous solutions. J. Biol. Chem. 258, 6360–6370.

Schmidt, D.G. and Payens, T.A.J. (1976) Surface and Colloid Science. Vol. 9. E. Matijevic', Ed. Wiley, New York, pp. 165–229.

Shinoda, K. (1963) in Colloidal Surfactants. B.I. Tamamushi and T. Isemura, Eds. Academic Press, New York, chap. 1.

Smith, R. and Tanford, C. (1972) The critical micelle concentration of L-α-dipalmitoylphosphatidylcholine in water and water/methanol solutions. J. Mol. Biol. 67, 75–83.

Swisher, R.D. (1987) Surfactant biodegradation. Marcel Dekker, New York.

Tausk, R.J.M., Karmiggelt, J., Oudshoorn, C. and Overbeek, J.Th.G. (1974a) Physical chemical studies of short-chain lecithin homologues. Part 1: Influence of the chain length of the fatty acid ester and of electrolytes on the critical micelle concentration. Biophys. Chem. *1*, 175–183.

Tausk, R.J.M., Oudshoorn, C. and Overbeek, J.Th.G. (1974b) Physical chemical studies of the short-chain lecithin homologues. Part 3: Phase separation and light scattering studies on aqueous dioctanoyl lecithin solutions. Biophys. Chem. *2*, 53–63.

Tausk, R.J.M., Van Esch, J., Karmiggelt, J., Voordouw, G. and Overbeek, J.Th.G. (1974c) Physical chemical studies of short-chain lecithin homologues. Part 2: Micellar weights of dihexanoyl- and diheptanoylecithin. Biophys. Chem. *1*, 184–203.

Timascheff, S.N. (1978) in Physical Aspects of Protein Interactions. N. Catsimpoolas, Ed. Elsevier–North Holland, Amsterdam, pp. 219–273.

van Voorst Vader, F. (1961) The pre-association of surfactant ions. Trans. Faraday Soc. *57*, 110–115.

Vrij, A. and Overbeek, J.Th.G. (1962) Scattering of light by charged colloidal particles in salt solutions. J. Coll. Sci. *17*, 570–588.

Wasylewski, Z. and Kozik, A. (1979) Protein–nonionic detergent interaction. Eur. J. Biochem. *95*, 121–126.

Wells, M.A. (1974) The mechanism of interfacial activation of phospholipase A_2. Biochem. *13*, 2248–2257.

CHAPTER 4

LIPID HYDRATED STATES AND PHASE BEHAVIOR

4.1. LYOTROPIC MESOMORPHISM

Phospholipids such as diacylphosphatidylcholine (lecithin) exhibit interesting behavior in the presence of water. In general, the phospholipids do not pass directly from the crystalline state to a solution in the presence of water. Various hydrated phases are encountered before solution of the phospholipids in water occurs. Such behavior is called lyotropic mesomorphism. The lyotropic phases also exhibit thermotropic mesomorphism; in other words, the particular phase obtained is a function both of water content and of temperature.

The importance of the thermotropic phase transition temperature can be seen when we appreciate that when water diffuses into the lattice it does so into the polar (ionic) region only when the temperature is reached at which the hydrocarbon chains "melt." If the temperature is higher than this, there is a simultaneous dissociation of the ionic lattice by the penetration of water and melting of the hydrocarbon chain region. The temperature of the transition (T_c line, see below) depends upon the nature of the hydrocarbon chains and on the polar region of the molecule, the amount of water present, and on any solutes dissolved in the water. Once the water has penetrated into the lattice of the amphiphile and the sample is then cooled to below the T_c line, the hydrocarbon chains rearrange themselves into an orderly crystalline lattice, but the water is not necessarily expelled from the system. These phases containing crystalline chain regions are sometimes called gels.

Bear et al. (1941) were the first to study, by X-ray diffraction, the lyotropic phases obtained by mixing natural lipids with water. Among the nerve lipid fractions studied was lecithin, which, on addition of two parts of water to one of lecithin, gives rise to a phase with a lamellar structure, liquidlike hydrocarbon chains, and a long spacing of 69 Å, some 60% greater than that of the dry material. For over a quarter of a century, no further work was published concerning either the structure or the range of existence of lecithin-water phases. In 1967 three papers by independent groups were published; two of the papers (Small, 1967; Reiss-Husson, 1967) were concerned with lecithin extracted from egg yolk, and the other (Chapman et al., 1967) covered a homologous series of 1,2-diacyl-*sn*-phosphatidylcholines (lecithins).

4.2. THE LAMELLAR PHASES

On addition of water, the transition temperature, T^*, of a phospholipid where the hydrocarbon chains melt is lowered to a limiting value, T_c. This transition temperature is the minimum temperature required for water to penetrate between the layers of the lipid molecules. Above the T_c line on the phase diagram, the phosphatidylcholine-water system exists in a mesomorphic lamellar phase, in which the hydrocarbon chains are in a melted condition. The composition of the system at maximum hydration is ~40 wt% water. Addition of more than 50 wt% water gives rise to a two-phase system consisting of fragments of the lamellar phase at maximum hydration dispersed in the excess water (see Fig. 4.1).

When the phosphatidylcholine-water system is cooled below the T_c line, the hydrocarbon chains adopt an ordered packing. The structure of this phase, the gel, is lamellar where the hydrocarbon chains are packed in various types of subcells.

The phase diagrams of the different chain-length lecithin-water systems are essentially equivalent and are disposed along the temperature axis according to the melting temperature (T_c) of the hydrocarbon chains. Differential scanning calorimetric curves of a 1,2-diacyl-*sn*-phosphatidylcholine-water system obtained over a range of water concentrations show that in the range $1.0 \geq c \geq 0.8$, the temperature of the endothermic transition T decreases steadily to a limiting value (T_c).

The cooperativity of the transition, as indicated by the narrowing of the melting range, also increases with increasing water. In this concentration range, despite the fact that an appreciable amount of water is present, no transition is observed in the heating or cooling curves due to any melting of ice or freezing of the water present. When the water content is greater than 20% ($c < 0.8$), the lipid endothermic transition temperature (T_c) remains constant and a peak at 0°C, due to the melting of ice, can now be observed in the heating curve. As the concentration of water in the mixture further increases, so does the size of this peak. Quantitative studies are interpreted as showing that a proportion of the water is bound to the lecithin in a fixed ratio of 1:4 by weight (10 moles water/mole lecithin). This bound water is due to the formation of a hydrate

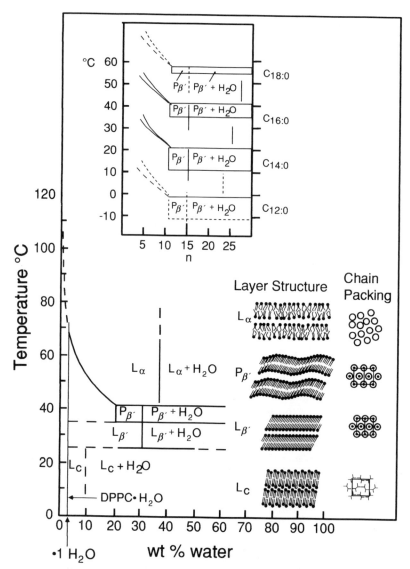

Fig. 4.1. Partial condensed phase diagrams of the dipalmitoylphosphatidylcholine-water system. The solid lines demarcate areas at which major molecular rearrangements occur (dashed where data were insufficient). Inset shows the comparison of phases for four different phosphatidylcholines as a function of the number of water molecules n associated with each phospholipid molecule. After Janiak et al. (1979).

structure associated with the polar group. The amount of bound water is independent of the fatty acid composition of the phospholipid, but is dependent upon the nature of the hydrophilic group.

All the different lecithins studied behave similarly. The heat absorbed at T_c for lecithin-water systems with c > 0.7 is chain-length dependent. The temperature interval between this peak and the main endothermic peak (T*) increases as the chain length of the lecithin becomes shorter.

The detailed packing arrangements of the gel structures—i.e., where the chains are in a crystalline form—have been studied. There is hexagonal, orthorhombic, and triclinic subcell hydrocarbon chain packing in these phases (Janiak et al., 1979). The condensed phase diagram up to approximately 170°C for dipalmitoyl lecithin is given in Figure 4.1. The insert shows the phase diagrams for the saturated phosphatidylcholines from C_{12} to C_{18} and, except for the increase in the chain-melting transition as the chain length increases and the increase in number of water molecules in L_α at maximum hydration, the characteristics are all similar. A composite diagram of calorimetric, volumetric, and X-ray data of hydrated DPPC is shown in Figure 4.2.

Recently an empirical relationship has been developed for the transition temperature T_c from the gel to the liquid-crystalline state of a series of fully hydrated saturated phosphatidylcholines. The expression is:

$$T_c = 154.2 + 2.0(\Delta C) - 142.8(\Delta C/CL) - 1512.5 (1/CL) \qquad (4.1)$$

ΔC is the effective chain-length difference in C—C bond lengths between the two acyl chains for the lipid in the gel state bilayer, and CL is the effective length of the longer of the two acyl chains, also in C—C bond lengths.

The empirical equation has been used to calculate the T_c value of bilayers of 163 molecular species of saturated phosphatidylcholines (Huang, 1991). Agreement between calculated and observed values are good, with 5.3% being the largest percent difference for the lipids examined. (The T_c values are only calculated for phosphatidylcholines with $\Delta C/CL$ values in the range of 0.09–0.40.)

The fluid character of the bilayer matrix of biomembranes is different from that of a simple paraffin melt. The ΔH values associated with the gel to liquid-crystal transition are lower than the ΔH values for the melting of pure hydrocarbons. The same holds true for the entropy change in the process. The incremental ΔS per CH_2 group is only about 1 e.u. for the gel to liquid-crystal transition of bilayers, but is almost twice as large for the melting of simple paraffins. These thermodynamic results show that the hydrocarbon chains in the bilayer core are not as disordered as they are in a pure liquid hydrocarbon.

The introduction of double bonds into the lipid hydrocarbon chain decreases the melting point, the enthalpy, and the entropy of the main chain transition, i.e., hexagonally packed chains to liquidlike chains. For example, the introduction of a single double bond at the 9–10 position of octadecanoic acid

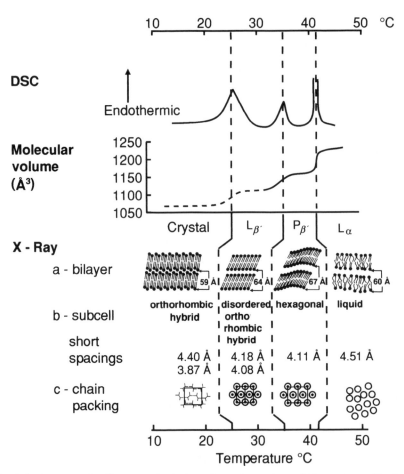

Fig. 4.2. A composite diagram of calorimetric, volumetric, and X-ray data of hydrated DPPC showing phases and transitions. After Janiak et al. (1979).

decreases the chain-melting point by 39°C for a single-chain substitution (e.g., distearoyl lecithin, 54°C; stearoyl-oleoyl lecithin, 15°C), while substitutions of both chains lower it even further (dioleoyl lecithin, −20°C). *Trans* double bonds have a lesser effect. The addition of a single double bond in one of the acyl chains lowers the melting temperature, and the addition of two double bonds in a single chain decreases the chain-melting point by about the same (70°C) as the addition of a single double bond in both chains. The addition of a third or fourth double bond has little or no further effect.

The position of the double bond appears to be a critical factor in determining the transition temperature and enthalpy of the lipid main endothermic phase transition (Huang et al., 1991). The shift of the double bond from the carbonyl

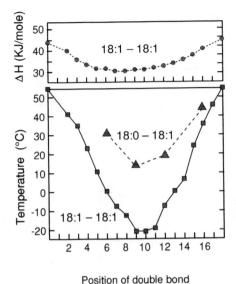

Fig. 4.3. Enthalpies (ΔH) and temperatures of the major thermal transition, in excess water for *cis*-unsaturated dioctadecenoic phosphatidylcholines (■) plotted against the Δ-position of unsaturation. Equivalent data for the dioctadecenoic phospholipid (●) are plotted for comparison. Transition temperatures of *sn*-1-octadecenoic phosphatidylcholines (▲) are also shown. From Barton and Gunstone (1975), with permission of the publisher.

region toward the terminal -CH$_3$ results in a progressive decrease in the chain-melting temperature from approximately 40°C to about −20°C, when the double bond is essentially in the midpoint of the fatty acid chain. The resulting enthalpy is also decreased in a regular fashion (see Fig. 4.3).

4.2.1. The Order Profile Parameter

In liquid crystals the degree of angular disorder of the molecules is expressed by the term *order parameter*. A set of order parameters have been defined by the equations

$$S_{ii} = \tfrac{1}{2}(\overline{3 \cos^2 \theta_i} - 1); \qquad i = 1, 2, 3, x, y, z \qquad (4.2)$$

where θ_i is the instantaneous angle between the molecular Cartesian coordinate i and the director z'. This angle varies with time due to fluctuations and $\cos^2 \theta_i$ is a time average for one molecule. The order parameter of the lipid chains has been studied by Seelig and Seelig (1974, 1980) and shows that, in the region of constant parameter, gauche conformations can occur only in complementary pairs, leaving the hydrocarbon chains essentially parallel to each other. Deuterium probe results differ from those obtained using spin-labeled molecules—e.g., the spin labels detect a continuous decrease of the order parameter, whereas the deuterium probe shows that the order parameter remains approximately constant for the first nine segments.

4.2.2. Correlation With Monolayer Properties

A correlation exists between the lipid monolayer properties at the air–water interface of lipids and the properties of the lipid bilayers in aqueous dispersions (Phillips and Chapman, 1968). The "condensed monolayer" corresponds to the crystalline or gel phase and the expanded state to the "fluid" or melted state that occurs above the lipid transition temperature. Similar thermotropic phase changes occur with the monolayers as occur with lipid bilayers. The isotherms observed at different temperatures with dipalmitoylphosphatidylcholine have been described in Chapter 2.

All monolayer states are possible with the saturated lecithin and phosphatidylethanolamine homologues (Albrecht et al., 1978). It is apparent that if the hydrocarbon chains are sufficiently long, condensed monolayers are formed, whereas with shorter chains liquid-expanded films occur. These two limiting states are sufficiently well defined so that at any particular temperature only one of the homologues studied exhibits the transition state. The data indicate that variations in hydrocarbon chain length that do not give rise to change in monolayer state do not have a significant effect on the π-A curves. Temperature changes can also give rise to the condensed and expanded states for a monolayer of a single homologue. Obviously, a sufficiently low temperature causes the film to become completely condensed, whereas at higher temperatures it is fully expanded. Monolayers in the two limiting states are more or less invariant with temperature, and it is the sensitivity of the phase transition to temperature that leads to the variety of isotherms (see Fig. 4.4a and 4.4b).

The molecules in a completely condensed phosphatidylethanolamine monolayer are much more closely packed than are those in the equivalent lecithin monolayer. This also correlates with the lipid bilayer behavior. The lecithins have a lower transition temperature in a bilayer structure than does the equivalent chain length phosphatidylethanolamine. This presumably arises for steric factors associated with the larger polar groups of the lecithin molecules.

4.2.3. Effects of Cholesterol

For some years it was known that cholesterol could affect and apparently condense monolayers (at the air–water interface) of certain unsaturated phospholipids. The meaning of this was, however, obscure and controversial; some workers believed that a *cis*-double bond was essential for condensation to occur. The invoked unusual structures and complexes between the lipid and cholesterol. A study of monolayer systems showed that phospholipids containing *trans*-double bonds, and even saturated phospholipids, could exhibit similar effects (Chapman et al., 1966).

Studies using deuterium nuclear magnetic resonance spectroscopy (NMR) were made with model biomembranes containing various amounts of cholesterol. Addition of cholesterol at the equimolar level (about 33 wt%) to the lipid results in an increase in quadrupole splitting from 3.6 to 7.8 kHz, corresponding to an increase in molecular order parameter (from $S_{mol} = 0.18$ to $S_{mol} = 0.41$). Cooling the sample to a temperature some 5°C below that of the gel to liquid-

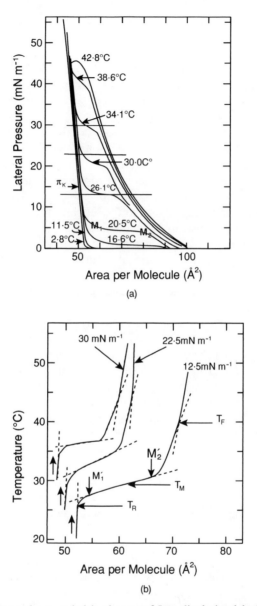

Fig. 4.4. **a.** Continuously recorded isotherms of L-α-dipalmitoyl lecithin (DPPC) on pure water substrate. Note the arrow at π_k, which indicates a break in the isotherm 2.8°C. **b.** Continuously recorded isobars of monomolecular film of L-α-DPPC. The curves were recorded at increasing temperatures. From Albrecht et al. (1978), with permission of the publisher.

crystal phase transition temperature ($T_c = 23\,°C$) has little effect on the quadrupole splitting, consistent with previous data. Cooling the pure lipid to the same temperature, however, results in hydrocarbon chain crystallization into the rigid crystalline gel phase, and a broad, rather featureless spectrum with $\Delta v_Q \sim 14.0$ kHz is observed. Analysis of this result in terms of a molecular order parameter is not possible, since the details of the motion of the rest of the hydrocarbon chain are unclear.

A feature of the inclusion of cholesterol into biomembrane structures is that the presence of large amounts of cholesterol prevents lipid chain crystallization and hence removes phase transition characteristics. With the 1,2-dipalmitoyl-*sn*-phosphatidylcholine (DPPC) cholesterol-water system, the thermotropic phase change in the presence of water occurs at a convenient temperature ($41\,°C$). The addition of cholesterol to the lecithin in water slightly lowers the transition temperature between the gel and the lamellar fluid crystalline phase, and markedly decreases the heat absorbed at the main transition. No transition is observed with an equimolar ratio of lecithin to cholesterol in water. This ratio corresponds to the maximum amount of cholesterol that can be introduced into the lipid bilayer before cholesterol precipitation takes place (see Ladbrooke and Chapman, 1969).

X-ray evidence indicates that a lamellar arrangement occurs and that at 50% cholesterol an additional long-spacing pattern occurs due to the separation of crystalline cholesterol. These results may be interpreted in terms of penetration of the lipid bilayer by cholesterol. In the lamellae of aqueous lecithin, the chains are hexagonally packed and titled at $58\,°C$. It can be envisaged that penetration will be facilitated when the chains are vertical. This causes an increase in the X-ray long spacing. At concentrations of cholesterol greater than 7.5%, the long spacing decreases. Above this critical concentration, a reduction occurs of the cohesive forces between the chains producing chain fluidization (Ladbrooke et al., 1968).

The effect of the presence of the cholesterol molecules is to modulate the lipid fluidity above and below the transition temperature of the lipid. Below the transition temperature the bilayer is fluidized, and above the transition temperature it is rigidified.

The relationship between the cholesterol content of a multilamellar lipid-water system and the permeability of the bilayer to water has been studied by Blok et al. (1977). These workers studied the osmotic shrinkage of the dipalmitoylphosphatidylcholine liposomes in glucose solutions. Below the transition temperature, the liposomes are relatively impermeable to water. Above the phase transition, there is a dramatic increase in water permeability. Addition of cholesterol leads to a progressive reduction in water permeability, which is associated with the ordering effect that cholesterol has upon the lipid chains.

4.3. PHASE SEPARATION

Most biomembranes contain a range of lipid classes and a variety of acyl chain lengths and degrees of unsaturation (see Chap. 1).

The first studies on phase separation of lipid-water systems were discussed by Ladbrooke and Chapman (1969), who reported studies of binary mixtures of lecithins using calorimetry. These authors examined mixtures of distearoyl and dipalmitoyl lecithin (DSL-DPL) and also distearoyl lecithin and dimyristoyl lecithin (DSL-DML). With the DSL-DPL mixtures, the phase diagram shows that a continuous series of solid solutions are formed below the T_c line. It was concluded that compound formation does not occur and that with this pair of molecules having only a small difference in chain length, cocrystallization takes place.

With the system DSL-DML, monotectic behavior was observed with limited solid solution formation. Here the difference in chain length is already too great for cocrystallization to occur so that as the system is cooled, migration of lecithin molecules occurs within the bilayer to give crystalline regions corresponding to the two compounds. The use of spin labels such as TEMPO (a phosphatidylcholine derivatized with 2,2,6,6-tetramethylpiperidine-1-oxyl, having a nitroxide group ($=N-O\cdot$) with a free electron) to examine phase separation of mixed lipid systems was reported by Shimshick and McConnell (1973) on similar lipid mixtures.

Metal ion interactions have been known for some years to affect the thermotropic phase transitions of soap systems. The thermotropic phase transition of stearic acid occurs at 114°C for the sodium salt and at 170°C for the potassium salt. These phase transitions can be linked to the monolayer characteristics. Similar effects are observed with certain phospholipids in the presence of divalent ions.

Early studies of stearic acid monolayers showed that interaction with Ca^{2+} ions caused an increase in surface pressure (i.e., condensation) and also decreased the permeability to water. The same effect has been observed with phosphatidylserine monolayers, but Na^+ and K^+ addition gave no such condensation. Later, more extensive studies showed that a variety of acidic phospholipid monolayers undergo an increase in surface potential and decrease in surface pressure on addition of Ca^{2+} and other bivalent cations. Phosphatidylserine is found to be more selective than phosphatidic acid, but for both systems the order of cation effectiveness is

$$Ca^{2+} > Ba^{2+} > Mg^{2+} \tag{4.3}$$

The formation of linear polymeric complexes was proposed to account for these findings.

Electric fields should also change the transition temperature T_c of a lipid phase transition if the dielectric constants (ϵ_1 and ϵ_2) in the two phases are different. The shift in transition temperature is given (Sackman 1979) by

$$\Delta T = \frac{T_c(\epsilon_2 - \epsilon_1)E^2}{\Delta H 8\pi\rho} \tag{4.4}$$

when ΔH is ca. 42 kJ mol^{-1}, $\rho = 1$ gcm^{-3}, $\epsilon_2 - \epsilon_1 \geq 10$, one expects ΔT

$\approx 0.1\,°C$ at a field strength $E = 10^5\ Vcm^{-1}$. This is a very small effect indeed and has not yet been observed.

Dramatic conformational changes may be effected in membranes of certain charged lipids by variations in the pH of the aqueous phase. Under favorable conditions (low ionic strength), phase changes may also be triggered by variations in the concentration of monovalent ions. Thus, for example, for phosphatidic acid, T_c varies abruptly between pH 8 and pH 10 and is lower by about 10°C in the twofold charged state. At low ionic strength, T_c is also a sensitive function of the salt concentration. For dimyristoylphosphatidylethanolamine (DMPE), T_c decreases with increasing ionic strength in the unchanged state (pH 4), whereas the reverse is valid for the fully charged state (pH 8.5).

4.4. HEXAGONAL PHASES

The hexagonal H_I phase of lipid-water systems was first found by Marsden and McBain (1948) in hydrated dodecylsulfonic acid (23–70% in water). Its structure was elucidated by Luzzati et al. (1958). The H_{II} phase of phospholipids was discovered, and its structure deduced by Luzzati and Husson (1962), in a lipid extract from human brain containing 52% PE, 35% PC, and 13% phosphoinositides (PI), at 37°C and at water contents below 22 wt%. The topology of normal H_I and inverse H_{II} phases is shown in figure 4.5.

An excellent review of the hexagonal phase has recently appeared (Seddon, 1990).

(a) H_I

(b) H_{II}

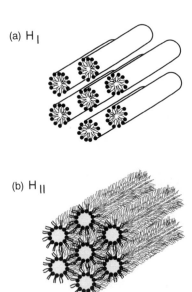

Fig. 4.5. Topology of normal (H_I) and inverse (H_{II}) hexagonal phases.

4.5. CUBIC PHASES

Cubic phases are the most complex among the mesomorphic structures occurring in lipid-water systems. A general structure consisting of interpenetrating rod networks was proposed in 1968 by Luzzati and coworkers. Thus the polar headgroups were suggested to form the rods and the hydrocarbon chains into a continuous matrix. The structure determination was performed on X-ray data from an anhydrous lipid, strontium myristate.

Cubic lipid-water phases (Q^{224} and Q^{230}) form remarkable structures with perfect long-range three-dimensional periodicity, although the molecules exhibit a dynamical disorder at atomic distances like those in liquids. Monoolein is an example of a lipid forming such a phase, which can contain up to 40% (w/w) of water. It is transparent and very viscous, and it can coexist in equilibrium with excess of water (with lipid monomer concentration as low as about 10^{-6} M). Due to the short-range disorder, these phases are called liquid crystals. The phase diagram of the system monoolein-water is shown in Figure 4.6.

The structure of these phases (Wieslander and Rilfors, 1977) has been described in terms of two three-dimensional networks of rods, mutually intertwined and unconnected (see Fig. 4.7): the rods, whose surface is lined by the polar headgroups of the lipid molecules, are filled by water and are embedded in the hydrocarbon matrix. A most remarkable property of these cubic structures is the presence of a unique hydrocarbon medium, continuous throughout the structure, and of two disjoined three-dimensional webs of water channels, each continuous throughout the structure, mutually interwoven and unconnected. These "minimal surfaces without intersections" are well known to mathematicians. The structure of the two cubic phases is topologically related to that of an isolated lipid bilayer: two disjoined and continuous water media are

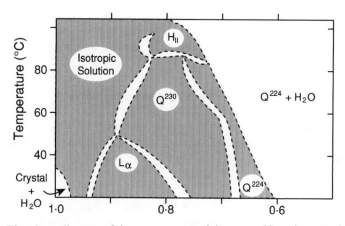

Fig. 4.6. The phase diagram of the system monoolein-water. Note the extended region over which the phase Q^{224} is found in equilibrium with excess water. From Hyde et al. (1954), with permission of the publisher.

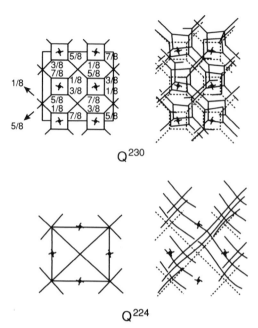

Fig. 4.7. The structure of the cubic phases. The two structures consist of two continuous three-dimensional networks of rods, mutually intertwined and unconnected. The rods are filled by water and are surrounded by the apolar medium containing the hydrocarbon chains and the unsubstituted glycerol headgroups. The thick lines represent the axes of the rods. Left frames, representation of the unit cell with the position of the axes of the rods and of some of the symmetry elements. Right frames, perspective view of the structure. Upper frames, phase Q^{230}. The rods are linked tetrahedrally four by four. From Hyde et al. (1954), with permission of the publisher.

present, separated from each other by a unique and continuous hydrocarbon septum. By contrast, in the lamellar phase the hydrocarbon and the water media are each subdivided into an infinite number of disjoined planar layers.

These and other cubic phases are commonly observed in lipids, but often at temperature and water content remote from physiological. Exceptions are the monoglycerides (Fig. 4.7), which adopt the cubic phases Q^{224} and Q^{230} in the high temperature–high water content region of the phase diagram.

If the cubic monoolein-water phase is shaken in a bile salt solution, a dispersion is formed with kinetic stability like a liposomal dispersion. It is sometimes possible to see, using a polarizing microscope, an outer birefringent layer with radial symmetry. The core is isotropic. These dispersions are formed in the region of the ternary system, where the cubic phase exists in equilibrium with water and the L_α-phase. The dispersion is obviously due to a localization of the L_α-phase outside cubic particles.

The cubic phase can also be dispersed by strongly amphiphilic proteins. Caseins, for example, which also are very effective as emulsifiers, can disperse the cubic phase like simple surfactants, such as bile salts. (See reviews by Larssen [1988] and Lindbloom and Rilfors [1989].)

REFERENCES AND FURTHER READING

Albrecht, O., Gruler, H. and Sackmann, E. (1978) Polymorphism of phospholipid monolayers. J. Phys. (Paris) *39*, 301–313.

Barton, P.G. and Gunstone, F.D. (1975) Hydrocarbon chain packing and molecular motion in phospholipid bilayers formed from unsaturated lecithins: Synthesis and properties of sixteen positional isomers of 1,2-dioctadecenoyl-*sn*-glycero-3-phosphocholine. J. Biol. Chem. *250*, 4470–4476.

Bear, R.S., Palmer, K.J. and Schmitt, F.O. (1941) X-ray diffraction studies of nerve lipids. J. Cell. Comp. Physiol. *17*, 355–367.

Blok, M.C., Van Deenen, L.L.M. and de Gier, J. (1977) The effect of cholesterol incorporation on the temperature dependence of water permeation through liposomal membranes prepared from phosphatidylcholines. Biochim. Biophys. Acta *464*, 509–518.

Chapman, D., Walker, D.A. and Owens, N.F. (1966) Physical studies of phospholipids. II. Monolayer studies of some synthetic 2,3-diacyl D.L. phosphatidylethanolamines and phosphatidylcholines containing trans double bonds. Biochim. Biophys. Acta *120*, 148–155.

Chapman, D., Williams, R.M. and Ladbrooke, B.D. (1967) Physical studies of phospholipids. Part 6: Thermotropic and lyotropic mesomorphism of some 1,2-dialcyl-phosphtidylcholines (lecithins). Chem. Phys. Lipids *1*, 445–475.

Huang, C. (1991) Empirical estimation of the gel to liquid-crystalline phase transition temperature for fully hydrated saturated phosphatidylcholines. Biochemistry *30*, 26–30.

Hyde, S.T., Andersson, S., Ericsson, B. and Larsson, K. (1984) A cubic structure consisting of a lipid blilayer forming an infinite periodic minimum surface of the gyroid type in the glycerolmonooleate-water system. Z. Krist. *168*, 213.

Janiak, M.J., Small, D.M. and Shipley, G.G. (1979) Temperature and compositional dependence of the structure of hydrated dimyristoyl lecithin. J. Biol. Chem. *254*, 6068–6078.

Ladbrooke, B.D., Williams, R.M. and Chapman, D. (1968) Studies on lecithin-cholesterol-water interactions by differential scanning calorimetry and X-ray diffraction. Biochim. Biophys. Acta *150*, 333–340.

Ladbrooke, B.D. and Chapman, D. (1969) Thermal analysis of lipids, proteins and biological membranes. Chem. Phys. Lipids *3*, 304–367.

Larssen, K. (1988) Anesthetic effect and a lipid bilayer transition involving periodic curvature. Langmuir *4*, 215.

Lindblom, G. and Rilfors, L. (1989) Cubic phases and isotropic structures formed by membrane lipid, possible biological relevance. Biochim. Biophys. Acta *988*, 221–256.

Luzzati, V. and Husson, F. (1962) The structure of the liquid crystalline phases of lipid-water systems. J. Cell Biol. *12*, 207–219.

Luzzati, V., Mustacchi, H. and Skoulios, A. (1958) The structure of the liquid-crystal phases of some soap + water systems. Discuss. Faraday Soc. *25*, 43–50.

Luzzati, V., Tardieu, A., Gulik-Krzywicki, T., Rivas, E., Reiss-Husson, R. (1968) Structure of the cubic phases of lipid-water systems. Nature *220*, 485–488.

Marsden, S.S. and McBain, J.W. (1948) X-ray diffraction in aqueous systems of dodecyl sulfonic acid. J. Am. Chem. Soc. *70*, 1973–1974.

Phillips, M.C. and Chapman, D. (1968) Monolayer characteristics of saturated 1,2-diacylphosphatidylcholine (lecithins) and phosphatidylethanolamines at the air-water interface. Biochim. Biophys. Acta *163*, 301–313.

Reiss-Husson, F. (1967) Structure des phases liquide-crystallines de differents phospholipides, monoglycerides, sphingolipides, anhydres ou en presence d'eau. J. Mol. Biol. *25*, 363–382.

Sackman, E. (1979). Light induced charge separation in biology and chemistry. E. Gerischer and J.J. Katz, Eds. Verlag Chemie, Weinheim.

Seddon, J.M. (1990) Structure of the inverted hexagonal (H_{II}) phase, and non-lamellar phase transitions of lipids. Biochim. Biophys. Acta *1031*, 1–69.

Seelig, A. and Seelig, J. (1974) The dynamic structure of fatty acyl chains in a phospholipid bilayer measured by deuterium magnetic resonance. Biochemistry *13*, 4839–4845.

Seelig, J. and Seelig, A. (1980) Lipid conformation in model membranes. Q. Rev. Biophys. *13*, 19–61.

Shimshick, E.J. and McConnell, H.M. (1973) Lateral phase separation in phospholipid membranes. Biochemistry *12*, 2351–2360.

Small, D.M. (1967) Observations on lecithin. Phase equilibria and structure of dry and hydrated egg lecithin. J. Lipid Res. *8*, 551–557.

Wieslander, Å. and Rilfors, L. (1977) Qualitative and quantitative variations of membrane lipid species in *Acholeplasma laidlawii*. Biochim. Biophys. Acta *466*, 336–346.

CHAPTER 5

THE LIPOSOMAL STATE

5.1. LIPOSOMES

The name liposome is used to describe vesicles where there is an aqueous volume enclosed by a membrane of lipid molecules. Liposomes form spontaneously when certain lipids—e.g., lecithins—are heated above a critical temperature in aqueous media. The vesicles can range in size from tens of nanometers to tens of microns in diameter. They can be made so that they entrap quantities of hydrophilic materials within their aqueous compartment and hydrophobic materials within the membrane. Liposomes can be made of natural constituents such that the liposome membrane forms a bilayer structure that is similar to the lipid portion of a natural cell membrane. (The introduction of charged lipids such as stearylamine or phosphatidic acid into the liposomes can increase their stability, inhibit aggregation, and increase the separation of multilayers.)

Liposomes can be formed by a variety of methods so as to control the size and also the number of "membranes," or bilayers, associated with the vesicles. The vesicles may have a single bilayer membrane or multiple concentric membrane lamellae. Liposomes of different sizes often require different methods of preparation and are classified for practical convenience according to their size.

5.2. THE FORMATION OF LIPOSOMES

5.2.1. Types of Liposomes

1. **Small unilamellar vesicles (SUVs).** These are liposomes at the lowest limit of size possible for phospholipid vesicles. This limit varies slightly

according to the ionic strength of the aqueous medium and the lipid composition of the membrane. It is about 25 nm for dipalmitoyllecithin liposomes.

2. **Multilamellar vesicles (MLVs).** These vesicles have a wide range of sizes (up to 10,000 nm); each vesicle consists of, say, five or more concentric lamellae. Vesicles composed of just a few concentric lamellae are sometimes called oligo-lamellar liposomes.

3. **Large unilamellar vesicles (LUVs).** These liposomes have diameters of the order of 50 to 10,000 nm.

4. **Intermediate-sized unilamellar vesicles (IUVs).** This term is sometimes used in the literature. These vesicles have diameters of the order of magnitude of 100 nm. A very convenient way of producing vesicles of the order of 100 nm diameter is by the extrusion of MLVs through filters of controlled pore size using high pressure (800 psi). This technique was developed by Hope et al. (1983), who defined such vesicles as VETs (vesicles by the extension technique).

Other terms used in the literature include REV (reverse-phase evaporation vesicle), DRV (dried-reconstituted vesicle) and MVL (multivesicular liposome). Fig. 5.1 shows a schematic representation of the main classes of liposomes and their size ranges. The encapsulated aqueous volumes range from 0.2–1.5 liters per mole of lipid for SUVs to 1–4 liters per mole of lipid for MLVs.

5.2.2. Methods of Preparation

Most methods of liposome preparation give a fairly heterogeneous population of vesicles with a wide distribution of sizes, particularly toward the lower end of the range. The lipid to entrapped volume ratio varies markedly with liposome size.

5.2.2.1. Multilamellar vesicles.
Multilamellar vesicles are produced merely by raising the temperature of the lipid above a critical temperature (T_c) with the lipid in contact with water (as shown by Bangham et al., 1965). Various slight modifications of this simple approach are available. For example, the lipid is dissolved in an organic solvent and dried down in a round-sided glass vessel of large volume so as to form a thin film. The temperature of drying down is usually recommended to be above T_c. After drying down, water is then added and multilamellar vesicles form spontaneously when the system is shaken.

5.2.2.2. Small unilamellar vesicles.
These vesicles were first produced by using ultrasonic irradiation to break up suspensions of MLV preparations (Huang, 1969). A suspension of multilamellar vesicles is taken and the vesicles are completely broken down in the process. Two common methods of sonication are carried out, using either a probe or a bath ultrasonic disintegrator.

Multilamellar vesicle (MLV)

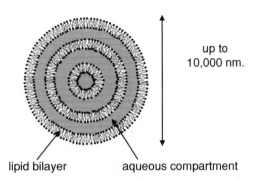

up to
10,000 nm.

lipid bilayer   aqueous compartment

Small unilamellar vesicle (SUV)

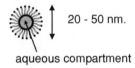

20 - 50 nm.

aqueous compartment

Large unilamellar vesicle (LUV)

aqueous
compartment

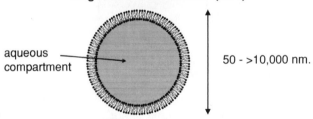

50 - >10,000 nm.

Fig. 5.1. The structure and dimensions of unilamellar (SUV and LUV) and multilamellar (MLV) liposomes.

The probe is employed for suspensions that require high energy in a small volume (e.g., high concentrations of lipids or a viscous aqueous phase), while the bath is more suitable for large volumes of dilute lipids. These sonicated vesicles give excellent high resolution ^{1}H NMR spectra, and studies have been made of the inner and outer layers of such vesicles using shift reagents such as lanthanide ions.

Various other methods have been devised to form lipid vesicles.

Ethanol injection. This method was originally reported by Batzri and Korn, 1973. An ethanol solution of lipids is injected rapidly into an excess of saline or other aqueous medium through a fine needle. The force of the injection is usually sufficient to achieve complete mixing, so that the ethanol is diluted almost instantaneously in water and phospholipid molecules are dispersed evenly throughout the medium. This procedure can yield a high proportion of small unilamellar vesicles (diameter ~ 250 Å [25 nm]).

Ether injection method. The ether injection method (Deamer and Bangham, 1976) involves injecting an immiscible organic solution into an aqueous phase through a narrow bore needle, at a temperature at which the organic solvent is removed by vaporization during the process. The slow vaporization of solvent gives rise to an ether:water gradient extending on both sides of the interfacial lipid monolayer, resulting in the eventual formation of a lipid bilayer sheet that folds in on itself to form a sealed vesicle.

Reverse-phase evaporation method. This method uses "water-in-oil" emulsions (Szoka and Papahadjopoulos, 1978). The process involves an emulsion that is the reverse of the standard "oil-in-water" emulsion. The novel step in the preparation is the removal of solvent from the emulsion by evaporation. The droplets are formed by bath sonication of a mixture of the two phases, and then the emulsion is dried down to a semisolid gel in a rotary evaporator under reduced pressure. The monolayers of phospholipid surrounding each water compartment are then closely opposed to each other. Next the gel is subjected to vigorous mechanical shaking with a vortex mixer, in order to bring about the collapse of a certain proportion of the water droplets. The aqueous content of the collapsed droplets provides the medium required for suspension of the newly formed liposomes. After conversion of the gel to a homogeneous free-flowing fluid, the suspension is dialyzed in order to remove the last traces of solvent.

5.2.3. The Stability of Liposomes

Small unilamellar vesicles (<400 Å [40 nm] in diameter) are prone to fusion, particularly at the lipid phase transition temperature. The permeability of liposome membranes depends very much on their composition, and also on the solute that is entrapped.

Liposomes have been shown to be permeable to water, ions, and nonelectrolytes, although the permeability depends upon the chemical composition of the liposome. Positively charged liposomes (e.g., lecithin plus a positively charged lipid such as stearylamine) are impermeable to cations, while negatively charged liposomes (e.g., those containing phosphatidic acid) are permeable to cations. The permeability of liposomes to protons is low. Anions diffuse rapidly through negatively and positively charged lipid membranes. Increasing the saturation or length of the phospholipid fatty acyl chains causes a decrease in the permeability of the liposomes to all solutes. This is a reflection of the T_c transition temperature of the lipids (see Chapter 4). General anesthetics such as ether and chloroform cause an increase in cation permeability but no increase in glucose permeability.

Liposomes have been prepared with archaebacterial lipids (Chapter 7). Freeze fracture studies show that the liposomes cross-fracture rather than through the midplane of the lipid bilayer, which is consistent with the formation of single lipid membranes rather than of the normal bilayer type (Chang et al., 1990).

Various methods have been employed to increase the stability of liposomes: (1) by introducing large amounts of cholesterol (molar/molar) into the liposomes, (2) by cross-linking membrane components covalently, (3) by using methods such as glutaraldehyde fixation, and (4) osmification or polymerization of alkyne-containing phospholipids—e.g., using diacetylene lipids followed by irradiation with ultraviolet light or γ-irradiation (Freeman and Chapman, 1988). One method that does not restrict the relative mobility of adjacent phospholipids is to incorporate long aliphatic branched-chain polymers (e.g. polyvinyl alcohols esterified with palmitic or stearic acid). These compounds, when incorporated up to about 10% by weight into phosphatidylcholine membranes, can substitute for cholesterol in reducing the leakage of medium-sized solutes. The incorporation of sphingomyelin into the liposomes can also increase their stability.

The ability of liposomes to retain their entrapped solvents may also be affected by freezing and by drying. When liposomes undergo dehydration, they often crack and release their entrapped solutes. Various added materials have been used to overcome this problem. The materials are similar to those used in cryoprotection of cells—e.g., pyrrolidone and various sugar molecules. One example of the latter is trehalose. This can produce a free-flowing dry liposome powder that retains its solute molecules, and the liposome can be reconstituted by adding water to the dry powder.

5.3. LIPOSOMES AS CARRIERS

Soon after the existence of phospholipid liposomes was described by Bangham and Horne (1962, 1964), it was realized that liposomes had potential as carriers of drugs and might be of value in numerous clinical situations (Sessa and Weissman, 1969; Gregoriadis et al., 1971). It is of interest to note that Bernard postulated the existence of liposomes in 1947, when discussing microscopic studies of myelinin figures formed in ammonium oleate–water systems. He clearly realized the significance of the unilamellar bilayer in the biological context.

The similarity of composition and structure of a phospholipid liposomal bilayer to that of the bilayer in a biological membrane suggested that liposomes would be "biocompatible" and hence ideal carriers for drugs in the body. Both water-soluble and oil-soluble drugs might be carried in the aqueous space and bilayer, respectively. The additional possibility of modifying the liposomal surface to target the liposome to a particular cell type by exploiting known characteristics of potential cell surface receptors suggested that a liposome was a manifestation of the "magic bullet" envisaged by Ehrlich (1906). The targeting of liposomes would seem to represent a major step toward the goal spelled out by Ehrlich in an address to the German Chemical Society in 1909 when he said, "We must learn to aim, and to aim in a chemical sense."

Much has been achieved in the field of liposomal drug targeting since the discovery of liposomes by Bangham, and a vast technology has been developed in the liposomal field (Gregoriadis and Allison, 1980; Gregoriadis, 1984; Knight, 1981; Ostro, 1987; New, 1990). Despite the very considerable research effort, relatively few liposomal drug preparations have reached the commercial stage, although many clinical trials are in progress. Table 5.1 lists some of the many applications of liposomes as delivery systems. These include enzyme replacement therapy; delivery of anticancer drugs, fungicides, bactericides, genetic material, and moisturizers and antiflammatory agents for skin; and radionuclides for tumor imaging. The number of potential applications of liposomal delivery is almost limitless, but at the present time the delivery of the fungicide amphotericin B and the anticancer anthracycline drugs such as doxorubicin (Adriamycin) are showing the most promise, up to the clinical development stage (Bangham, 1992).

Amphotericin B is a monocyclic polyene antifungal antibiotic produced by *Streptomyces nodosus*. It is effective against systemic fungal infections, particularly candida and aspergillis species, which infect immunocompromised patients with acquired immune deficiency syndrome (AIDS). The fungicide is toxic and interacts with cell membrane sterols, causing channels which allow the cellular contents to leak (Gray and Morgan, 1991). The chemical side effects of amphotericin B include fever, chills, disturbance of the electrolyte balance, and impairment of renal function. By encapsulating the drug in a phospholipid liposome, the toxic effects are believed to be reduced. In the commercial material, AmBisome, the amphotericin B is encapsulated in a liposome consisting of hydrogenated soya phosphatidylcholine, cholesterol, distearoylphosphatidylglycerol and α-tocopherol. The effectiveness of this delivery system probably in part relates to the fact that injected liposomes are rapidly cleared from the blood by the reticuloendothelial system (RES) (mononuclear phagocyte system [MPS]) of the liver, spleen, and bone marrow, which are the main sites of systemic fungal infections. The treatment of the parasitic infection of the liver, Leishmaniasis, by liposomal preparations of amphotericin B and 8-aminoquinolines is for similar reasons also effective.

The drug doxorubicin, which inhibits cell division and is an effective anticancer drug, also has side effects, the most serious being progressive and irreversible damage to the heart. Liposomally encapsulated doxorubicin, while being as effective as the free drug, is significantly less toxic (Leyland-Jones, 1993). Liposomally encapsulated doxorubicin may also have an added advantage: encapsulation has been shown to modulate drug resistance in cultured cells (Thierry et al., 1993). Multidrug resistance is a major obstacle in cancer treatment. Drug resistance in cultured tumor cells is often due to the expression of a plasma membrane P-glycoprotein, encoded by multidrug resistance genes, that functions as a drug-efflux pump. Liposomally encapsulated doxorubicin has been shown to be more toxic than the free drug to several cancer cell lines, including those of the colon, ovaries, and breast.

The efficiency of liposomes as carriers will not only depend on the effec-

TABLE 5.1. Applications of Liposomes as Delivery Systems

Agents to be Delivered in Liposome	Examples	Clinical or Scientific Objective
Lysosomal enzymes[a]	Hexosaminidase A	Tay-Sachs disease (enzyme replacement therapy)
	Glucocerebrosidase, peroxidase	Gauchers disease (enzyme replacement therapy)
Cytosolic enzymes[a]	Hypoxanthine guanine phosphoribosyltransfrase	Lesch Nyhan syndrome (enzyme replacement therapy)
Anticancer drugs[b]	Duanorubicin, doxorubicin (Adriamycin), epirubicin, methotrexate	Inhibiting tumor growth and control of metastasis
Vaccines (protein antigens)[c]	Malaria merozoite, malaria sporozoite, hepatitis B antigen, rabies virus glycoprotein	Liposome-induced improved immune response, acts as adjuvant
Fungicides[d]	Amphotericin B	Systemic fungal infections, particularly candida and aspergillus species in immunocompromised (AIDS) patients
Antileishmanial agents[e]	8-Aminoquinolines, amphoterincin B	Treatment of leishmaniasis (leishmanias are parasites that infect the reticuloendothelial system)
Radionuclide[f]	In-111, Tc-99m	Diagnostic imaging of tumors
Genetic material[g]	Lipochromosomes, DNA	DNA transfection, genetic treatment of inherited disorders
Antibacterial agents	Triclosan[h]	Oral hygiene, delivery of bactericide to oral bacteria
	Clindamycin hydrochloride[i]	Treatment of acne vulgaris
Moisturizers, antiinflammatory[j] agents	Sodium pyrrolidone carboxylate	Cosmetic moisturizer
	Triamcinolone (fluorohydroxy-hydrocortisone)	Antiinflammatory and antiallergic agent
Antiviral drugs[k]	3'-azido-3'-deoxythymidine (AZT)	HIV (AIDS) therapy

Data from: [a]Finkelstein and Weissmann (1978); [b]Leyland-Jones (1993); [c]Alving (1987); [d]Chopra et al. (1992); [e]Gray and Morgan (1991); [f]Williams et al. (1984); [g]Fraley et al. (1981); Mannino and Gould-Fogerite (1988); [h]Jones et al. (1993); [i]Skalko et al. (1992); [j]Strauss (1989); [k]Phillips (1992).

tiveness of the agent being carried but on the success of directing the liposome to the required site (targeting) and the way in which it interacts with the site and transports the agent into the target biosurface (plasma membrane). These two aspects of liposomal delivery systems are discussed below.

5.4. LIPOSOMAL TARGETING

There are several methods of targeting liposomes to cellular systems and other biosurfaces (e.g., skin, hair, and teeth). These can be classified under the headings of natural or passive targeting, physical targeting, compartmental targeting, and active or ligand-mediated targeting.

5.4.1. Natural Targeting (Passive Targeting)

Natural targeting exploits the body's natural defense mechanism against foreign particles in the bloodstream. The phagocytes of the liver and spleen that form part of the reticuloendothelial system rapidly clear liposomes from the circulation, and hence in the treatment of infections and carcinomas in these organs natural targeting can be exploited. There has been considerable discussion about the cells responsible for the uptake of liposomes by the liver—both parenchymal and Kupffer cells have been found to preferentially take up small liposomes (50 nm diameter) (Scherphof et al., 1983). The size distribution of the fenestrations in the endothelia determines the extent of vesicle uptake by parenchymal cells. It has also been suggested that while the initial site of uptake may be the Kupffer cells, a secondary translocation process occurs that results in the transfer of lipid to parenchymal cells. The surface charge of the liposomes plays an important part in the uptake of liposomes in perfused rat liver in the absence of blood; increasing the negative charge on the liposomes decreases the rate of uptake, while positively charged liposomes are taken up more quickly the higher the charge (Nicholas and Jones, 1986), although blood modifies this behavior as a consequence of interactions between blood components with the liposomes (see sec. 5.6) (Nicholas and Jones, 1991). Natural targeting can clearly be exploited in the treatment of infections of the RES such as leishmaniasis, brucellosis and listerosis, but methods of circumventing the RES for drug delivery to other organs by liposomes are required.

5.4.2. Physical Targeting

Physical targeting depends on directing a liposome to a particular location where the environment is manipulated to induce the liposome to release its contents. Two common types of physical targeting involve the use of "temperature-sensitive" liposomes and "pH-sensitive" liposomes. In temperature-sensitive targeting, our ability to manipulate the physical state of the liposomal bilayer by choice of lipid chain length, and hence chain-melting temperature, is utilized (Weinstein et al., 1979). If the chain-melting temperature of the liposomal lipid is controlled to be above body temperature, the encapsulated agent will only leak out slowly at body temperature; if the target site is heated above the chain-melting temperature of the liposomal bilayers, the agent will be released. The advantages of temperature-sensitive liposomes are that extra-vascular targets can be treated and the liposomes do not need to specifically bind to the target cell. However, their use requires a knowledge of the location of the site to be treated. Significant therapeutic effects have been observed in the treatment of tumors with adriamycin using temperature-sensitive liposomes (Ono et al., 1992).

Sites of infection and inflammation and also tumors have a lower pH than healthy tissue, and this can be exploited in the development of pH-sensitive liposomes, which release their encapsulated contents when they pass into a region of low pH. When cells take up liposomes by receptor-mediated endo-cytosis, the liposomes enter endosomes (the pre-lyosomal compartment) that have an acidic pH (5.0–6.5) that is sufficiently low to destabilize pH-sensitive liposomes. Liposomes that are pH sensitive can be prepared from phospha-tidylethanolamines (PE) in combination with fatty acids or other substances, such as cholesterol hemisuccinate or palmitoylhomocysteine, that confer a charge on the liposomes at neutral pH. PE liposomes become unstable when uncharged due to the formation of the inverted hexagonal phase, so that when the liposome charge is reduced at low pH, the liposome disrupts with concom-itant release of its contents (Bentz et al., 1987).

Light-sensitivity may also be described as a means of physical targeting. By choice of photolabile lipids, it is possible to make liposomes that will release their entrapped contents on exposure to light of appropriate wavelengths. There are several ways of making liposomes that will leak their contents when irra-diated: (1) by exploiting light-induced phase transitions in the bilayer lipids; (2) by use of lipids that undergo photodegradation; and (3) by use of lipids that undergo photoisomerization (Pidgeon and Hunt, 1987). For example, the phospholipid derivatives of retinoic acid such as 1,2-diretinoyl-*sn*-glycero-3-phosphocholine (DRPC) form liposomes that are impermeable in the dark,

DRPC

but irradiation induces conformational changes (geometric isomerization in the retinoid groups together with lipid degradation) and the liposomes become permeable. The use of magnetic fields to direct magnetized liposomes encapsulating magnetite and iron particles has also been investigated as a means of physical targeting (Kiwada et al., 1986).

5.4.3. Compartmental Targeting

This method of targeting involves the administration of the liposomes at sites where the therapeutic effect is required. Liposomes administered subcutaneously or intravenously enter the lymphatic nodes that drain the injected tissue and will eventually localize in the liver and spleen. Delivery to the lungs can be effected using a nebulizer or by intratracheal administration. Such methods are particularly appropriate in the treatment of neonatal respiratory distress syndrome—which is a consequence of a deficiency of pulmonary surfactant (largely phospholipid) (Avery and Mead, 1959)—and for the delivery of antiasthma drugs (Taylor and Newton, 1992), while arthritis can be treated by intra-articular administration (Knight et al., 1985).

The delivery of insulin by the oral route has been studied (Patel and Ryman, 1976) but without success. Liposomes are unstable in the gastrointestinal tract due to attack by bile salts and phospholipases, making oral delivery of liposomally encapsulated insulin perhaps unlikely to be able to work without the development of more robust liposome formulations such as polymerized or partially polymerized liposomes.

5.4.4. Ligand-Mediated Targeting (Active Targeting)

The most sophisticated method of liposome targeting involves the modification of the surface of liposomes by the introduction of methods that will bind specifically to receptor sites on the surface of the target cell. The surfaces of cells are generally coated with carbohydrate residues covalently linked to the glycolipids and glycoproteins of the plasma membrane. The plasma membranes of mammalian cells have also been found to contain so-called animal lectins (Drukamer, 1988; Lis and Sharon, 1991), which are receptors for oligosaccharide moieties. Thus ligand-mediated targeting of liposomes generally involves carbohydrate-protein interactions. Table 5.2 summarizes some of the types of ligand-mediated liposomal targeting. The materials used to modify the liposomal surface include glycolipids, glycoproteins, and antibodies raised to cell surface antigens (usually oligosaccharides), as well as lipopolysaccharides and polysaccharides. The spike glycoproteins of viruses mediate the interaction of the virus with the target cell surface receptors. The interaction results in the transfer of the virus genome into the host cell either by an injection mechanism, in the case of nonenveloped viruses, or by membrane fission or fusion, in the case of enveloped viruses. Virus spike glycoproteins can thus be used as a means of targeting liposomes to cells. Such liposomes are termed "virosomes."

TABLE 5.2. Ligand-Mediated Liposomal Targeting

Liposomes	Target Site
Glycolipid-bearing liposomes[a]	Carbohydrate receptors on cell surfaces
Glycoprotein-bearing liposomes[b]	Carbohydrate receptors on cell surfaces
Viral spike glycoprotein-bearing liposomes (virosomes)[c]	Cell surface virus receptors
Antibody-bearing liposomes (immunoliposomes)[d]	Cell surface antigens
Lectin-bearing liposomes[e]	Cell surface carbohydrates
Lipopolysaccharide- and polysaccharide-bearing liposomes[f]	Carbohydrate receptors on cell surfaces

Data from: [a]Jones (1994); [b]Jones (1994); [c]Helenuis et al. (1987); [d]Peeters et al. (1987), Wright and Huang (1989); [e]Sato and Sunamoto (1992), Jones (1994); [f]Sunamoto and Iwamoto (1987), Sato and Sunamoto (1992).

The attachment of antibodies to the liposomal surface to form "immunoliposomes" is another method of targeting that is highly specific, especially if monoclonal antibodies raised to cell surface antigens are used. A somewhat less specific method of targeting is to attach plant lectins that have specificity for particular monosaccharide residues which may be present on cell surfaces. Thus concanavalin A—which has a specificity for glucose and mannose residues and their pyranosides—and wheat germ agglutinin—which has a specificity for N-acetylneuraminic acid and N-acetylglucosamine—have been used for targeting to bacteria (Hutchinson et al., 1989; Jones et al., 1993). The incorporation of glycoprotein and glycolipids into the liposomal surface is relatively straightforward in that they are molecules of membrane origin and will naturally anchor into the liposomal bilayer during liposome preparation. In contrast, attaching proteins (antibodies and lectins) onto the liposomal surface to form "proteoliposomes" requires chemical methods and presents a characterization problem to determine the protein density on the liposomal surface, which will ultimately determine the effectiveness of targeting. There have been very considerable variations in the protein surface densities for different proteins reported in the literature (Hutchinson et al. 1989).

5.4.4.1. Preparation and characterization of proteoliposomes.
There are a variety of chemical methods that can be used for covalently linking proteins to liposomal surfaces (Martin et al., 1990). Most methods involve the derivatization of the protein to be attached and the activation of a lipid to be incorporated into the liposome, most commonly the activation of a PE. As examples we will consider two particular commonly used methods. The first

method involves the use of the "double agent" N-succinimidylpyridylthiopro-
pionate (SPDP). This will react with –NH$_2$ groups in both the lipid (PE) and

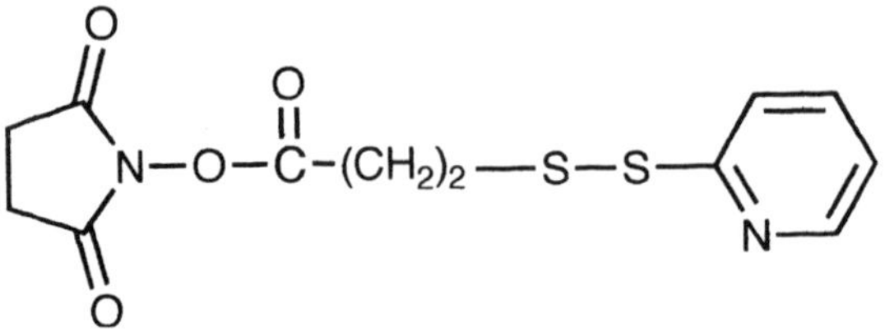

SPDP

protein at pH 7–8 in aqueous solutions to form the following derivatives plus
N-hydroxysuccinimide (NHS):

$$PE(NH_2) + SPDP \rightarrow PE\text{-}(NHCO(CH_2)_2\text{-}S\text{-}S\text{-}Pyr) + NHS$$

$$Prot(NH_2) + SPDP \rightarrow Prot\text{-}(NHCO(CH_2)_2\text{-}S\text{-}S\text{-}Pyr) + NHS$$

The derivatized PE can be added to the lipid mixture required to make the
liposomes. The protein derivative can be reduced with dithiothreitol to give
sulphydryl groups on the protein, i.e., Prot-(NHCO(CH$_2$)$_2$-SH). The reduced
protein derivative will then undergo an exchange reaction with the PE derivative
in the liposome to give the protein conjugated liposome

$$Liposome\ (PENHCO(CH_2)_2\text{-}S\text{-}S\text{-}(CH_2)_2\ CO\ HNProt)$$

Because the PE and protein derivatives are in different states of oxidation with
respect to the sulphur, there is negligible chance of the formation of cross-
linking between protein–protein and liposome–liposome. It is, however, important that dithiothreitol is rigorously removed from the reduced protein derivative, as its presence during conjugation would lead to unwanted cross-
linking.

A second method depends on using different double agents for derivatizing
the lipid (PE) and protein. The PE is derivatized with m-maleimidobenzoyl-
N-hydroxysuccinimide (MBS)

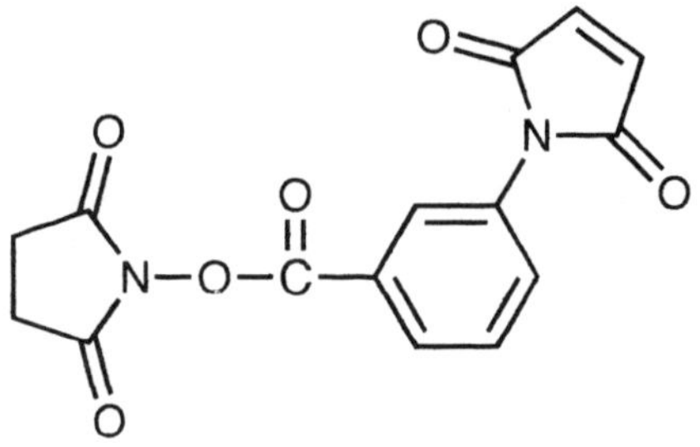

MBS

to yield

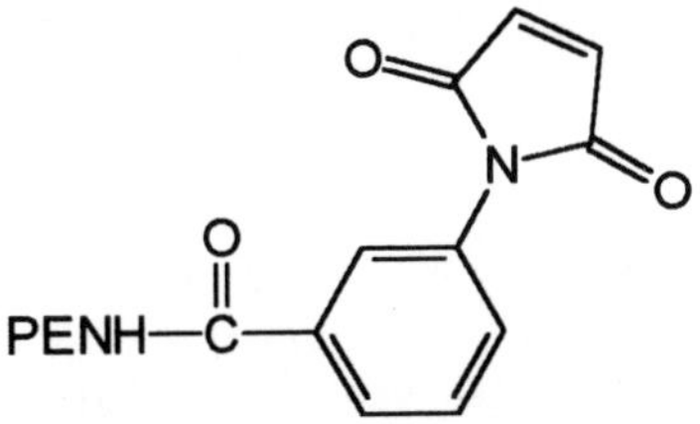

$+ NHS$

which is incorporated into the liposome at the preparative stage. The protein is derivatized with N-succinimidyl-S-acetylthioacetate

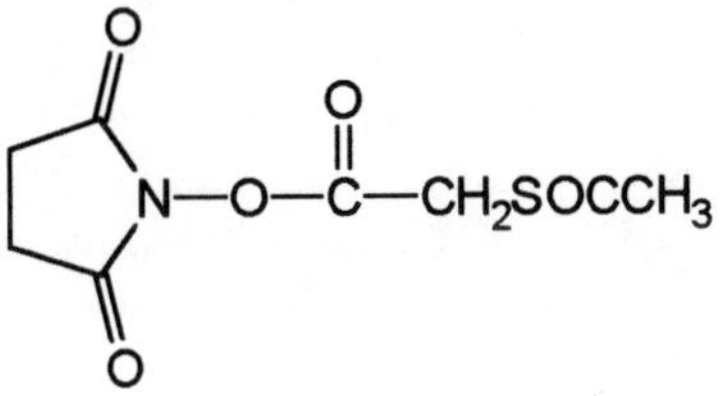

SATA

to yield

$$Prot - NH - CO\,CH_2\,S\,OCCH_3 + NHS$$

The protein derivative can then be "activated" by hydroxylamine (NH_2OH) to remove the acetyl groups, leaving sulphydryls which will undergo an addition reaction to the maleimido group of the derivatized PE

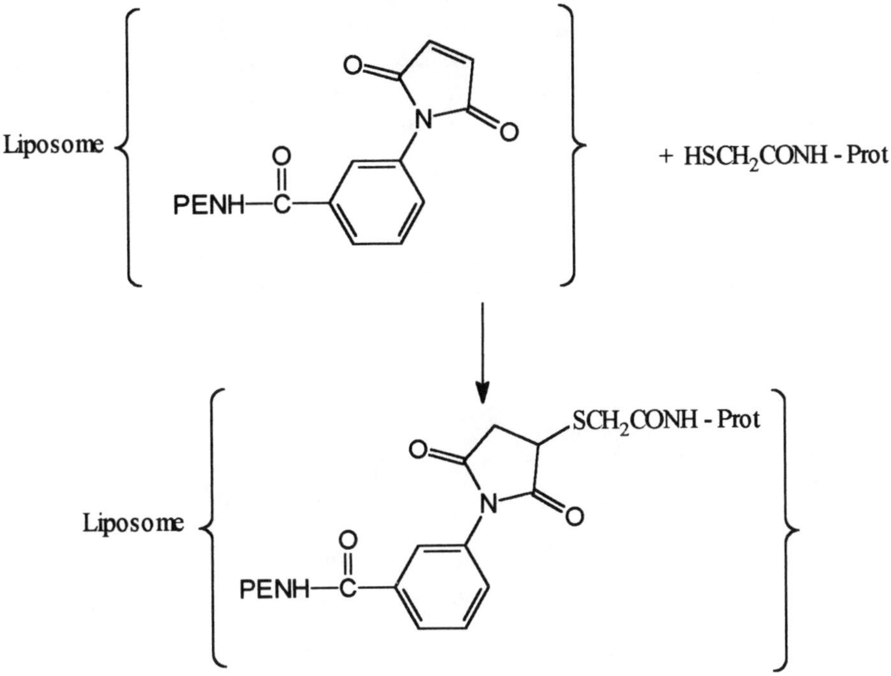

Liposome { ... } $+ HSCH_2CONH - Prot$

Liposome { ... }

Both of these methods involve reactions that will go easily at room temperature.

There are a number of factors that control the extent of protein conjugation to liposomes using the above methods. The extent of derivatization of the protein will depend on the number of lysine NH_2 groups on the protein surface that will react with the double agent. In case of reaction with SATA, the number of derivatized lysyl residues is generally small, not usually exceeding on average three or four for lectins and antibodies; however, this extent of derivatization is satisfactory for conjugation and should not lead to any major conformational changes to the protein structure. It should be borne in mind that, assuming a Poisson distribution for the derivatized species, an average of circa four derivatized lysyl residues means that approximately 70% of the protein molecules will have two to five derivatized groups and only approximately 19% the proteins will have precisely four derivatized groups.

The degree of conjugation of the liposomes can be controlled by the extent of incorporation of the "activated" lipid (the derivatized PE) in the liposome preparation; however, the amount of the PE derivative that is actually used in conjugation appears to be only a small percent of that incorporated. For lectin conjugation, a linear relationship is found between the mass of lectin conjugated per mole of liposomal lipid and the mole % PE derivative in the liposomes, as shown in Figure 5.2 for both SUV and REV liposomes covering size (weight average diameter, $\bar{d}_w$) ranges of 65–150 nm and 160–240 nm, respectively. Interestingly, the slopes of the plots are, within experimental error, the same for both types of liposome. The identity of the slopes suggests that the extent of protein conjugating is not influenced by the size of the liposomes but only by the amount of reactive lipid that controls the surface density of conjugated protein. This enables us to define a parameter for the average number of protein molecules per liposome. As liposome sizes are conveniently determined by dynamic light scattering, which gives a weight-average diameter ($\bar{d}_w$), it is appropriate to define a weight-average number of protein molecules per liposome ($\bar{P}_w$) according to the equation (Hutchinson et al., 1989)

$$\bar{P}_w = \frac{\sum_i P_i w_i}{\sum_i w_i} \qquad (5.1)$$

where P_i and w_i are the number of proteins per liposome and weight of liposomes of species i, respectively. For liposomes of a given size, $\bar{P}_w$ increases linearly at low mole % of reactive lipid (Fig. 5.3). However, for both concanavalin A and wheat germ agglutinin-conjugated liposomes, $\bar{P}_w$ approaches limiting values of the order of 40 nm^2 per molecule for SUVs and 100 nm^2 per molecule for REVs (Fig. 5.4). These figures suggest that the density of the lectins on the liposomal surface approaches a limit determined by packing considerations. From the dimensions of concanavalin A and wheat germ agglutinin, the projected areas—allowing for an exclusion effect—lie in the ranges

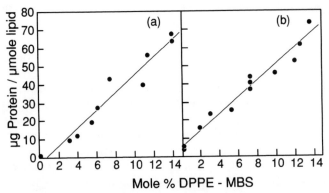

Fig. 5.2. Relationship between the extent of conjugation of wheat germ agglutinin (μg protein per μ mol lipid) to DPPC/PI (9:1, by wt.) liposomes as a function of DPPE-MBS (reactive lipid) incorporation. **a.** SUV. **b.** REV. From Hutchinson et al. (1989), with permission of the publisher.

of 52–200 nm^2 for concanavalin A and 52–112 nm^2 for wheat germ agglutinin, depending on the orientation of the molecules on the surface. The lectins appear to pack more tightly on the surface of SUVs than of REVs, which is consistent with the higher curvature of SUVs (Francis et al., 1992).

The interaction between liposomes carry site-directing molecules and their

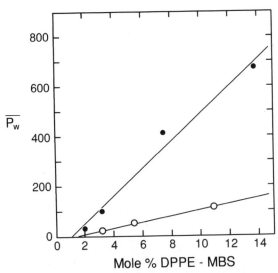

Fig. 5.3. Weight-average number of proteins (WGA) per liposome (DPPC/PI 9:1, by wt.) as a function of DPPE-MBS incorporation. $\bullet$, REV ($\bar{d}_w = 163 \pm 6$ nm); $\bigcirc$, SUV ($\bar{d}_w = 83 \pm 4$ nm). From Hutchinson et al. (1989), with permission of the publisher.

target site would be expected to be influenced by the concentration of the liposomes and the number of site-directing molecules on their surface ($\overline{P}_w$). The concentration (number per unit volume) of proteoliposomes is directly proportional to the lipid concentration (c) and inversely proportional to their surface area $4\pi(\overline{d}_w^2/2)^2$; i.e., the larger the liposomes the more lipid they contain. Targeting might be expected to increase with the parameter $(c/\overline{d}_w^2)\overline{P}_w$. Figure 5.5 shows a plot based on the above argument for the targeting of concanavalin A–bearing liposomes to a biofilm of the oral bacteria *Steptococcus mutans* as measured by inhibition of an enzyme-linked immunosorbent assay (ELISA) for antigens on the bacterium surface. Inhibition (and hence targeting) increases initially with $(c/\overline{d}_w^2)\overline{P}_w$ but approaches a limiting value, possibly because the binding of the protoliposomes to the biosurface is not as strong as the antibody used in the ELISA.

The preparation and characterization of immunoliposomes show some features that differ from those of lectin-bearing proteoliposomes. Some antibodies appear to adsorb nonspecifically on the surface of liposomes in the absence of activated lipid, and the relationship between the protein to lipid ratio and level

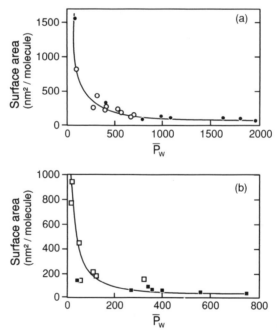

Fig. 5.4. a. Surface area of protein (area per molecule) on proteoliposomes (REV) conjugated with concanavalin A (con A) (●) and wheat germ agglutinin (WGA) (○) as a function of weight-average number of proteins per proteoliposome ($\overline{P}_w$). **b.** Surface area of protein (area per molecule) on proteoliposomes (SUV) conjugated with sConA (■) and WGA (□) as a function of weight-average number of proteins per proteoliposome ($\overline{P}_w$). From Francis et al. (1992), with permission of the publisher.

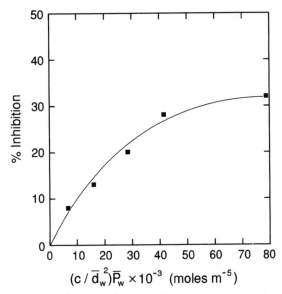

Fig. 5.5. Dependence of inhibition of ELISA for bacterial cell surface antigens on the parameter $(c/\bar{d}_w^2)\bar{P}_w$. SUV with surface-bound succinylated concanavalin A targeted to *Streptococcus mutans* biofilms. Adapted from Jones et al. (1993).

of reactive lipid is not linear as for lectin-bearing proteoliposomes (Fig. 5.2). Nonspecific I_gG adsorption to the surface of liposomes was noted by Senior et al. (1986). In fact, in some instances, such as in the conjugation of the antibody raised to placental alkaline phosphate (H17E2), the area per antibody on the liposomal surface due to nonspecific binding is smaller (i.e., surface density is higher) than when the antibody is conjugated using the PEMBS/antibody-SATA derivative chemistry, as described above, for both small (VETs $\bar{d}_w$ ~ 100 nm) and large (REV, $\bar{d}_w$ 200–330 nm) proteoliposomes (Hudson and Jones, 1993; Jones and Hudson, 1993). It would appear that on conjugation the antibody occupies a larger area on the liposomal surface when it is covalently attached than when it is physically adsorbed by hydrophobic interactions. The targeting of these immunoliposomes to tumor cells follows a similar pattern to the targeting of lectin-bearing proteoliposomes; targeting increases linearly with the parameter $\bar{d}_w^2/\bar{P}_w$ at constant proteoliposome concentration (Fig. 5.6).

   In general, the technology is available for the preparation of a very wide range of proteoliposomes for targeting to particular biosurfaces and many systems have been studied, including numerous antibodies (Peeters et al., 1987; Wright and Huang, 1989) and other site-directing macromolecules (Sato and Sunamoto, 1992; Jones, 1994). It is, however, important to rigorously characterize the resulting proteoliposomes in terms of not only their size distribution but the distribution of covalently linked site-directing molecules on the liposomal surface, because it is this parameter that plays a key role in determining

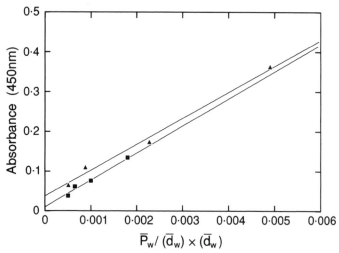

Fig. 5.6. Dependence of ELISA signal for antibody (H17E2) raised to placental alkaline phosphatase on immunoliposomes targeted to immobilized placental alkaline phosphatase (PLAP) on theparameter $\bar{P}_w/\bar{d}_w^2$. The immunoliposome lipid concentration was constant at 0.3 mM. ▲, REVs; ■, VETs. From Jones and Hudson (1993), with permission of the publisher.

targeting efficiency. In the future, one would expect that more emphasis will be placed on developing methods to produce antibody epitopes for targeting. In principle, it is only the antibody binding site (epitope) that is required, and this could be produced by genetic engineering (Verhoeyen and Riechmann, 1988). Such genetically engineered site-directing antibody fragments should lead to immunoliposomes with large surface densities of site-directing groups and improved targeting efficiency.

5.5. THE INTERACTION OF LIPOSOMES WITH CELLS

As sections 5.3 and 5.4 show, it is possible to attach site-directing molecules to the surface of liposomes and to produce systems targeted to a particular cell type. How effective a liposomal carrier is in the delivery of a particular therapeutic agent will depend on how the target cell receives the liposomes and processes its cargo. There are four principle ways in which liposomes interact with cells, as depicted in Figure 5.7; these are adsorption, endocytosis, lipid exchange, and fusion (Weinstein, 1981). The adsorption of liposomes to cell surfaces will depend on the surface characteristics of the liposomes; in the absence of site-directing molecules the charge on the liposomal surface will play an important part in the interaction. Most cells carry a negative charge due to anionic groups (e.g., carboxylate, sulphate) in their glycocalyx, and hence negatively charged liposomes would not be expected to adsorb readily unless mediated by bridging divalent metal ions. Other important physical

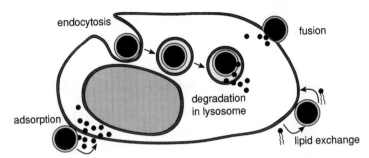

Fig. 5.7. Interaction of liposomes with a mammalian cell. From Lasic (1992), reprinted by permission of *American Scientist*, journal of Sigma Xi, the Scientific Research Society.

factors that might determine adsorption include hydrogen bonding, the hydrophobic interactions, and dispersion forces. Clearly liposomes with site-directing groups should adsorb. If only adsorption occurs, then the delivery of a drug will depend on the drug passively diffusing from liposome to cell down a concentration gradient. The liposome thus acts only to present a high drug concentration at the cell surface. For oil-soluble drugs, the possibility exists of direct transfer from the liposomal bilayer to the plasma membrane, assuming adsorption is sufficiently strong so that the liposome and cell bilayers are brought into close contact.

If endocytosis occurs, the entire liposome will be engulfed by the cell; however only a few cell types will endocytose liposomes (Ostro and Cullis, 1989). Monocytes, macrophages, and other white cells derived from the bone marrow effectively endocytose liposomes. Circulating monocytes and macrophages are part of the reticuloendothelial system (RES), of which the liver, spleen, lymph nodes, bone marrow, and lungs are the major organs. When liposomes adsorb to white cells (monocytes and macrophages), they are engulfed (Fig. 5.7) by the invaginated plasma membrane in what are called phagosomes, which fuse with cell lysosomes where the liposomal bilayers are disrupted; the liposomal lipid of biological origin may be reused and the liposomal contents released into the cell cytoplasm. Proteins of proteoliposomes will be degraded by lysosomal enzymes.

Lipid exchange involves the transfer of liposomal lipid into the plasma membrane of the cell. If a drug is covalently linked to a liposomal lipid, then it could become part of the plasma membrane bilayer. If the covalent linkage was labile in the cell cytoplasm, then this would result in the transfer of the drug from liposome to cell.

Ideally, the delivery of drugs to cells would be most satisfactory if the liposome fused with the plasma membrane, so that the drug was released directly into the cytoplasm and hence might avoid damaging effects of lysosomal enzymes, as occurs in endocytosis. This mechanism has been described as the Holy Grail of liposome engineering (Weinstein, 1981). Unfortunately, it has become increasingly clear that fusion between liposomes and cells is a relatively rare event, although there have been claims to the contrary (Huang, 1983).

5.6. THE PROBLEM OF THE RES IN DRUG DELIVERY BY LIPOSOMES

Following the demonstration that intravenously injected liposomes are rapidly removed from the circulation by the RES (Gregoriadis and Ryman, 1972a, 1972b) it became clear that the targeting of liposomes to cells other than those which form part of the RES requires a means of avoiding the RES.

The uptake of particles by the cells of the RES is preceded by the adsorption of substances from the blood. Wright and Douglas (1903) studied the role of blood in phagocytosis and they first showed that serum effected the phagocytosis of pathogenic bacteria. The materials adsorbed from serum which affect the phagocytosis of particles are called opsonins from the Greek *opson* (a substance added to food to give it flavor). Thus when liposomes are introduced into the circulation, they adsorb serum proteins from the blood which are thought to determine their recognition by mononuclear phagocytes (Patel, 1992). Serum components that retard or inhibit phagocytosis are called dysopsonins.

In order to avoid liposomes being taken up by the RES, various strategies have been adopted. One strategy is to saturate the RES with liposomes targeted to it by incorporation of 6-aminomannitol or 6-amminomannose, which have been shown to be preferentially taken up by the liver and spleen, followed by the liposomes encapsulating the drug (Proffitt et al., 1983). These procedures are examples of the use of a ''blockade'' to prevent the drug-carrying liposomes from being removed from the circulation and hence to increase the probability of the drug-carrying liposomes reaching the desired target cells such as tumors. Suppression of liposome uptake by the liver has also been brought about by the use of dextran sulphate (Patel et al., 1983), which also acts as a blockade.

An alternative approach to blocking the RES is to modify the liposomal surface so that the liposome is not recognized by phagocytic cells. This was first achieved by the incorporation of the ganglioside GM_1 into egg phosphatidylcholine-cholesterol liposomes (Allen and Chonn, 1987). Gangliosides are a class of glycophingolipids containing one or more N-acetylneuraminic acids (NANA) (sialic acids). GM_1 has the structure

$$R_1\text{-CH=CH-CH-OH}$$
$$|$$
$$R_2\text{CONHCH}$$
$$|$$
$$CH_2\text{-O - glc - gal - glcNAc - gal}$$
$$|$$
$$NANA$$

where R_1 and R_2 are alkyl chains.

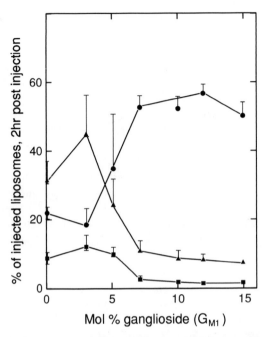

Fig. 5.8. Tissue distribution of Stealth liposomes containing entrapped ^{125}I-tyraminyl-inulin, 2 h postinjection, as a function of ganglioside concentration. Large unilamellar vesicles LUV (0.17 μm) of SM : PC, 4 : 1 containing GM$_1$: liver (▲), spleen (■), blood (●). From Allen and Chonn (1987), with permission of the publisher.

Fig. 5.8 shows the effect of incorporation of GM$_1$ into egg PC–sphingo-myelin liposomes with entrapped radioiodine (^{125}I-tyraminylinulin) on the tissue distribution of radioactivity after the liposomes have been injected into mice. Increasing the mole % GM$_1$ reduces the uptake by the liver and spleen and increases the blood level. Both surface sialic acid and bilayer viscosity have a synergistic effect on increasing the circulation times of liposomes. A bilayer viscosity intermediate between the liquid-crystalline and gel states results in optimal circulation times. Because of their property of being able to circulate in the blood without being apprehended by the RES, the GM$_1$ containing li-posomes have bene called Stealth liposomes (after the Stealth bomber, which is not detectable by radar) (Allen, 1989). In having sialic acids on their surface, Stealth liposomes to some degree resemble the surface of red blood cells, where the sialic acid is carried by a class of glycoproteins called glycophorins.

A further type of liposome designed to have reduced uptake by the RES has water-soluble polymers covalently linked to the surface, most commonly poly-ethyleneglycol (PEG) with a molecular mass from 1,000–5,000. The stability of these liposomes in the circulation is attributed to the steric barrier created by the surface-bound PEG and they have been called sterically stabilized li-posomes (SSL). SSL can be prepared by conjugating the PEG to activated PE;

several chemical methods have been described (Woodle and Lasic, 1992). The choice of lipid in the liposomal bilayer is not limited, and prolonged circulation is independent of cholesterol content, the degree of hydrocarbon chain saturation, liposomal lipid charge, and liposome size. The origin of steric stabilization is far from clear-cut, as Woodle and Lasic discuss (1992). PEG is a highly soluble polymer, and the PEG chains would rather interact with water molecules than with the surface of the liposome. It is suggested that they extend out from the surface of the liposome to form a "brush." The interaction of SSL with the glycocalyx of a cell will be unfavorable on entropic grounds due to the loss of conformational degrees of freedom when the PEG overlaps with the cellular surface polymers (entropic stabilization); on the other hand, interpenetration of the polymers might lead to energetically unfavorable contacts (enthalpic stabilization). In some conflict with stabilization conferred by the PEG, it is well established that high surface densities (50 mole % of lipid molecules carrying PEG) of low molecular mass PEG (degree of polymerization of about 15) leads to liposome fusion with cells. The presence of a very water soluble coating on the liposomes may inhibit the binding of opsonins, which in combination with the steric barrier created by the PEG and the increased hydrophilicity of the liposomal surface protects the liposome from uptake by the RES.

The RES is not the only obstacle presented to liposomes *in vivo*; on interaction with blood, a range of possible interactions with blood plasma proteins and lipoproteins is possible, as summarized in Fig. 5.9 (Bonté and Juliano, 1986; Senior, 1987). These interactions range from physical adsorption of proteins to the liposomal surface (the opsonins and dysopsonins are examples of this type of interaction) to penetration of more hydrophobic proteins into the bilayer, e.g., apolipoprotein subunits. However, the most important type of interaction, as far as the stability of the liposome is concerned, is that involving the exchange of liposome lipid between the liposomal bilayer and lipoproteins in the blood. All lipoproteins probably exchange lipid to some degree, but it is the exchange with high-density lipoprotein (HDL) that is generally believed

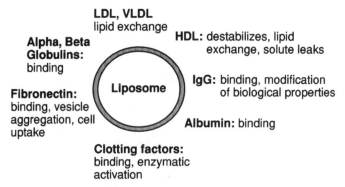

Fig. 5.9. Interactions of liposomes with blood plasma components. From Bonté and Juliano (1986), with permission of the publisher.

to lead to liposome instability in the blood. The instability of liposomes in the blood is to a certain extent controllable by suitable choice of lipid composition. Liposome stability can be increased by the use of saturated lipids with chain-melting temperatures above body temperature and by the inclusion of 3β-hydroxysterols such as cholesterol. The adsorption of serum opsonins has been reported to be inhibited in liposomes composed of sphingomyelin or saturated phospholipids. Such liposomes may adsorb dysopsonins more readily, so that they will not only be more stable in the blood but will have a reduced uptake by the RES (Moghimi and Patel, 1989).

REFERENCES AND FURTHER READING

Allen, T.M. (1989) in Liposomes in the Therapy of Infectious Diseases and Cancer. G. Lopez-Berestein and I.J. Fidler, Eds. Alan R. Liss, New York, pp. 405–415.

Allen, T.M. and Chonn, A. (1987) Large unilamellar liposomes with low uptake into the reticuloendothelial system. FEBS Lett. *223*, 42–46.

Alving, C.R. (1987) in Liposomes: From Biophysics to Therapeutics. M.J. Ostro, Ed. Marcel Dekker, New York.

Avery, M.E. and Mead, J. (1959) Surface properties in relation to atelectasis and hyline membrane disease. Am. J. Dis. Child. *97*, 917–923.

Bangham, A.D. (Dec. 15, 1992) Liposomes: Realizing their promise. Hospital Practice, pp. 51–62.

Bangham, A.D. and Horne, R.W. (1962) Action of saponin on biological membranes. Nature *196*, 952–953.

Bangham, A.D. and Horne, R.W. (1964) Negative staining of phospholipids and their structural modification by surface active agents as observed in the electron microscope. J. Mol. Biol. *8*, 660–668.

Batzri, S. and Korn, E.D. (1973) Single bilayer liposomes prepared without sonication. Biochim. Biophys. Acta *298*, 1015–1019.

Bentz, J., Ellens, H. and Szoka, F.C. (1987) Destabilization of phosphatidyletholamine-containing liposomes: Hexagonal phase and asymmetric membranes. Biochemistry *26*, 2105–2116.

Bernard, A.T. (1947) Note sur les figures "myéliniques" et leur interpretation physico-chimique. Arch. Roum. Pathol. Exp. Microbiol. *14*, 53–68.

Bonté, F. and Juliano, R.L. (1986) Interactions of liposomes with serum proteins. Chem. Phys. Lipids *40*, 359–372.

Chang, E.L., Rudolph, A. and Lo, S.L. (1990) A liposome-forming archaebacterial lipid. Biophys. J. *57*, 220a.

Chopra, R., Fielding, A. and Goldstone, A.H. (1992) Successful treatment of fungal infections in neutroperic patients with liposomal amphotericin (AmBisome): A report on 40 cases from a single centre. Leukaemia and Lymphoma *7*, 73–77.

Deamer, D.W. and Bangham, A.D. (1976) Large volume liposomes by an ether vaporization method. Biochim. Biophys. Acta *443*, 629–634.

Drukamer, K. (1988) Two distinct classes of carbohydrate-recognition domains in animal lectins. J. Biol. Chem. *263*, 9557–9560.

Ehrlich, P. (1906). Collected Studies on Immunology. Vol. 2. John Wiley & Sons, New York.

Finkelstein, M. and Weissman, G. (1978) The introduction of enzymes into cells by means of liposomes. J. Lipid Res. *19*, 289–303.

Francis, S.E., Hutchinson, F.J., Lyle, I.G. and Jones, M.N. (1992) The control of protein surface concentration on proteoliposomes. Colloids Surfs. *62*, 177–184.

Fraley, R., Straubinger, R.M., Rule, G., Springer, E.L. and Papahadjopoulos, D. (1981) Liposome mediated delivery of deoxyribonucleic acid to cells. Biochemistry *20*, 6978–6987.

Freeman, F.J. and Chapman, D. (1988) Polyinvisible liposomes as drug carriers. G. Gregoriadis, Ed.

Gray, A. and Morgan, J. (1991) Liposomes in Haematology. Blood Rev. *5*, 258–272.

Gregoriadis, G., ed. (1984) Liposome Technology. Vols. 1–3. CRC Press, Florida.

Gregoriadis, G. and Allison, A.C., Eds. (1980) Liposomes in Biological Systems. Wiley Interscience, New York.

Gregoriadis, G. and Ryman, B.E. (1972a) Lysosomal localization of B-fructofuranos-idase-containing liposomes injected into rats. Biochem. J. *129*, 123–133.

Gregoriadis, G. and Ryman, B.E. (1972b) Fate of protein-containing liposomes injected into rats: An approach to the treatment of storage diseases. Eur. J. Biochem. *24*, 485–491.

Gregoriadis, G., Leathwood, P.D. and Ryman, B.E. (1971) Enzyme entrapment in liposomes. FEBS Lett. *14*, 95–99.

Helenius, A., Doxsey, S. and Mellman, I. (1987) Viruses as tools in drug delivery. Ann. N.Y. Acad. Sci. (R.L. Juliano, Ed.) *507*, 1–6.

Hope, M.J., Bally, M.B., Webb, G. and Cullis, P.R. (1985) Production of large unilamellar vesicles by a rapid extrusion procedure. Characterisation of size distribution, trapped volume and ability to maintain a membrane potential. Biochim. Biophys. Acta *812*, 55–65.

Huang, C.H. (1969) Studies on phosphatidylcholine vesicles: Formation and physical characteristics. Biochemistry *8*, 334–352.

Huang, L. (1987) in Liposomes. M.J. Ostro, Ed. Marcel Dekker, New York, pp. 81–124.

Hudson, M.J.H. and Jones, M.N. (1993a) The preparation and characterisation of immunoliposomes with surface-bound placental alkaline phosphatase antibody. Colloids Surf. B: Biointerfaces *1*, 157–166.

Hutchinson, F.J., Francis, S.E., Lyle, I.G. and Jones, M.N. (1989) The characterisation of liposomes with covalently attached proteins. Biochim. Biophys. Acta *978*, 17–24.

Jones, M.N. (1994) Carbohydrate-mediated liposomal targeting and drug delivery. Adv. Drug Deliv. Rev., *13*, 215–250.

Jones, M.N. and Hudson, M.J.H. (1993b) The targeting of immunoliposomes to tumour cells (A431) and the effects of encapsulated methotrexate. Biochim. Biophys. Acta *1152*, 231–242.

Jones, M.N., Francis, S.E., Hutchinson, F.J., Handley, P.S. and Lyle, I.G. (1993) Targeting and delivery of bactericide to adsorbed oral bacteria by use of proteoliposomes. Biochim. Biophys. Acta *1147*, 251–261.

Kiwada, H., Sato, J., Yamada, S. and Kato, Y. (1986) Feasibility of magnetic liposomes as a targeting device for drugs. Chem. Pharm. Bull. (Tokyo) *34*, 4253–4258.

Knight, C.G., Ed. (1981) Liposomes: From Physical Structure to Therapeutic Applications. Elsevier, Amsterdam.

Knight, C.G., Bard, D.R. and Thomas, D.P.P. (1985) Liposomes as carriers of antiarthritic agents. Ann. N.Y. Acad. Sci. *446*, 415–428.

Lasic, D. (1992) Liposomes. Am. Sci. *80*, 20–31.

Leyland-Jones, B. (1993) Targeted drug delivery. Semin. Oncol. *20*, 12–17.

Lis, H. and Sharon, N. (1991) Lectin-carbohydrate interactions. Current Opinion in Structural Biology *1*, 741–749.

Mannino, R.J. and Gould-Fogerite, S. (1988) Liposome mediated gene transfer. Biotechniques *6*, 682–690.

Martin, F.J., Heath, T.D. and New, R.C.C. (1990) in Liposomes: A Practical Approach. R.C.C. New, Ed. IRL Press, Oxford, chap. 4.

Moghimi, S.M. and Patel, H.M. (1989) Serum opsonins and phagocytosis of saturated liposomes. Biochim. Biophys. Acta *984*, 384–387.

New, R.C.C., Ed. (1990) Liposomes: A Practical Approach. IRL Press, Oxford.

Nicholas, A.R. and Jones, M.N. (1986) The adsorption of phospholipid vesicles by perfused rat liver depends on vesicle surface charge. Biochim. Biophys. Acta *860*, 105–111.

Nicholas, A.R. and Jones, M.N. (1991) The effect of blood on the uptake of liposomal lipid by perfused rat liver. Biochim. Biophys. Acta *1074*, 105–114.

Ono, A., Eeno, M., Zoa, I. and Horikoshi, I. (1992) Basic study on hepatic artery chemoembolization using temperature-sensitive liposomes. J. Pharmacobio. Dyn. *15*, s-71.

Ostro, M.J., Ed. (1987) Liposomes: From Biophysics to Therapeutics. Marcel Dekker, New York.

Ostro, M.J. and Cullis, P.R. (1989) Use of liposomes as injectable-drug delivery systems. Am. J. Hosp. Pharm. *46*, 1576–1587.

Patel, H.M. (1992) Serum opsonins and liposomes: Their interaction and opsonophagocytosis. Crit. Rev. Ther. Drug Carrier Systems *9*, 39–90.

Patel, H.M. and Ryman, B.E. (1976) Oral administration of insulin by encapsulation within liposomes. FEBS Lett. *62*, 60–63.

Patel, K.R., Li, H.P. and Baldeschwieler, J.D. (1983) Suppression of liver uptake of liposomes by dextran sulfate 500. Proc. Natl. Acad. Sci. USA *80*, 6518–6522.

Peeters, P.A.M., Storm, G. and Crommelin, D.J.A. (1987) Immuniliposomes *in vivo*: State of the art. Adv. Drug Deliv. Rev. *1*, 249–266.

Phillips, N.C. (1992) Liposomal carriers for the treatment of acquired immune deficiency syndromes. Bull. Inst. Pasteur *90*, 205–230.

Pidgeon, C. and Hunt, C.A. (1987) Photolabile liposomes and carriers. Methods in enzymology *149*, 99–111.

Proffitt, R.T., Williams, L.E., Presant, C.A., Tin, G.W., Uliana, J.A., Gamble, R.C. and Baldeschwieler, J.D. (1983) Liposomal blockade of the reticuloendothelial system: Improved tumour imaging with small unilamellar vesicles. Science *220*, 502–505.

Sato, T. and Sunamoto, J. (1992) Recent aspects in the use of liposomes in biotechnology and medicine. Prog. Lipid Res. *31*, 345–372.

Scherphof, G.L., Roerdink, F., Dijkstra, J., Ellens, H., De Zanger, R. and Wisse, E. (1983) Uptake of liposomes by rat and mouse hepatocytes and Kupffer cells. Biol. Cell *47*, 47–58.

Senior, J.H. (1987) Fate and behaviour of liposomes *in vivo*: A review of controlling factors. CRC Crit. Rev. Therapeutic Drug Carrier Systems *3*, 123–193.

Senior, J., Waters, J.A. and Gregoriadis, G. (1986) Antibody-coated liposomes. The role of non-specific adsorption. Febs Letters *196*, 54–58.

Sessa, G. and Weissman, G.J. (1969) Formation of artificial lysosome *in vitro* J. Clin. Invest. *48*, 76a–77a.

Škalko, N., Čajkova, M. and Jalšenjak, I. (1992) Liposomes with clindamycin hydrochloride in the therapy of *Acne vulgaris*. Int. J. Pharmaceutics *85*, 97–101.

Strauss, G. (1989) Liposomes: From theoretical model to cosmetic tool. J. Soc. Cosmet. Chem. *40*, 51–60.

Sunamoto, J. and Iwamoto, K. (1987) Protein-coated and polysaccharide-coated liposomes as drug carriers. CRC Crit Rev. Ther. Drug Car. Sys. *2*, 117–136.

Szoka, F. and Papahadjopoulos, D. (1978) Procedure for preparation of liposomes with larger internal aqueous space and high capture by reverse-phase evaporation. Proc. Nat. Acad. Sci. USA *75*, 4194–4198.

Taylor, K.M.G. and Newton, J.M. (1992) Liposomes for controlled delivery of drugs to the lung. Thorax *47*, 257–259.

Thierry, A.R., Vigé, D., Coughlin, S.S., Belli, J.A., Dritschilo, A. and Rahman, A. (1993) Modulation of doxoribicin resistance in multidrug-resistant cells by liposomes. FASEB J. *7*, 572–579.

Verhoeyen, M. and Riechmann, L. (1988) Engineering of antibodies. BioEssays *8*, 74–78.

Weinstein, J.N. (1981) Liposomes as "targeted" drug carriers: A physical chemical perspective. Pure and Appl. Chem. *53*, 2241–2254.

Weinstein, J.N., Magin, R.L., Yatvin, M.B. and Zaharko, D.S. (1979) Liposomes and local hyperthermia: Selective delivery of methotrexate to heated tumours. Science *204*, 188–191.

Williams, L.E., Proffitt, R.T. and Lovisatti, L. (1984) Possible applications of phospholipid vesicles (liposomes) in diagnostic radiology. J. Nucl. Med. Allied Sci. *28*, 35–45.

Woodle, M.C. and Lasic, D.D. (1992) Sterically stabilized liposomes. Biochim. Biophys. Acta *1113*, 171–199.

Wright, A.E. and Douglas, S.R. (1903) An experimental investigation of the role of blood fluid in connection with phagocytosis. Proc. R. Soc. Ser. B. Biol. Sci. *72*, 357–365.

Wright, S. and Huang, L. (1989) Antibody-directed liposomes as drug delivery vehicles. Adv. Drug Deliv. Rev. *3*, 343–389.

CHAPTER 6

SURFACTANT INTERACTIONS WITH BILAYERS, MEMBRANES, AND PROTEINS

6.1. SURFACTANT-MEMBRANE INTERACTIONS

The extensive use of surfactants as solubilizing agents for membrane components, particularly functional proteins and glycoproteins, demonstrates the importance of surfactants in the field of preparative biochemistry. In the area of analytical biochemistry, polyacrylamide gel electrophoresis (PAGE) in the presence of sodium n-dodecylsulphate (SDS) is one of the most commonly used methods for analyzing the polypeptide composition of mixtures of proteins and glycoproteins and for estimating molecular weights of single-chain proteins and protein subunits (Weber and Osborn, 1975). The mechanism of interaction of surfactants with membranes and their constituents is of considerable interest. It is only from a complete understanding of the processes that occur when surfactants interact with cell membranes that the full potential of surfactant interactions can be exploited for the development of the most useful and appropriate preparative and analytical methods for solving particular biochemical problems. The wide range of commercially available surfactants means that the researcher is faced with a major problem in deciding which type of surfactant is appropriate for a specific task. It is only a knowledge of the fundamental aspects of membrane solubilization that enables correct decisions to be made with regard to the choice of surfactant for a particular problem. The amphipathic nature of surfactants, as discussed in Chapter 1, which lead to micellization (Chap. 3), is central to their role as membrane-solubilizing agents and use in reconstitution studies (Chap. 8). The essential concept that should be appreciated in the study of surfactant interactions in the biochemical context is that we are always concerned with multiple equilibria between the free surfactant monomer and the other species present, specifically surfactant-lipid mixed micelles, surfactant-lipid-macromolecule complexes, surfactant-macromolecule complexes and surfactant micelles. The composition of a solubilized system may thus be complex and the relative proportions of the different species present will be dependent on the ratios of surfactant to lipid and surfactant to macromolecule in the system, as well as on the physical factors such as ionic strength, temperature, and the surfactant chemistry, which determine the critical micelle concentration (cmc) and hence monomer concentration. The relative strength of the hydrophobic interactions between the surfactant, lipid and macromolecular membrane components will determine the composition of the system, and in this respect it is logical to consider systems in order of increasing complexity.

6.1.1. Surfactant-Lipid Interactions

Surfactants and lipids have the same overall amphipathic type of structure but differ considerably in their hydrophobic-lyophilic balance (HLB), which manifests itself in the concentration of monomeric species that can exist in aqueous solution. The cmcs of the most commonly used surfactants are in a range of the order of 0.1 to 30 mM, while naturally occurring lipids have vanishingly small cmcs of the order of 10^{-10}–10^{-13} M (see Chapter 3). The interaction of surfactants with lipids at most practical concentrations will generally be concerned with the interaction of the monomeric surfactant with some type of lipid mesophase (see Chapter 4). As many lipid lyotropic mesophases are based on a bilayer structure, it is appropriate to consider the interaction of a surfactant with a phospholipid bilayer. Figure 6.1 shows a schematic view of the events occurring as the effective molar ratio of surfactant to bilayer lipid, defined by the equation (Lichtenberg et al., 1983; Lichtenberg, 1985)

$$R_e = \frac{[S] - [S]_{monomer}}{[PL]} \tag{6.1}$$

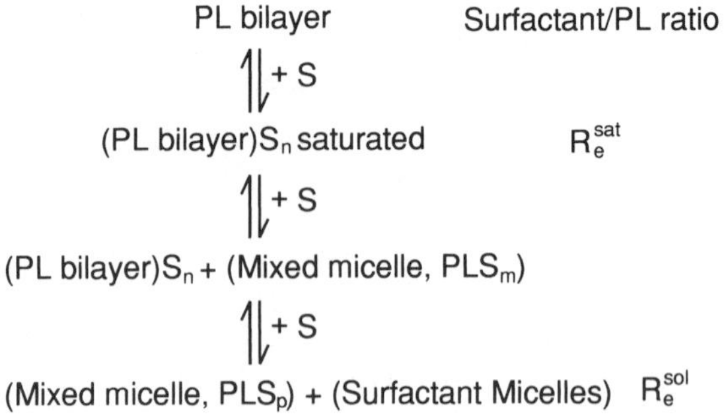

PL bilayer   Surfactant/PL ratio

$\Big\uparrow\Big\downarrow$ + S

(PL bilayer)S_n saturated R_e^{sat}

$\Big\uparrow\Big\downarrow$ + S

(PL bilayer)S_n + (Mixed micelle, PLS$_m$)

$\Big\uparrow\Big\downarrow$ + S

(Mixed micelle, PLS$_p$) + (Surfactant Micelles) R_e^{sol}

Fig. 6.1. Schematic representation of the sequence of events arising on exposure of a phospholipid bilayer (PL) to increasing amounts of surfactant (S). From Jones (1992), with permission of the Royal Society of Chemistry.

is increased, where [S] and [S]$_{monomer}$ are the molar concentrations of total and monomeric surfactant, respectively, and [PL] in the molar concentration of bilayer phospholipid. As R_e is increased, the monomeric surfactant progressively penetrates the bilayer until it becomes saturated at a ratio R_e^{sat}. Further increase in the surfactant concentration results in the formation of an increasing concentration of mixed surfactant/phospholipid micelles formed from the dissolution of the bilayer and terminating in the complete solubilization of the bilayer into mixed micelles at R_e^{sol}. Once the bilayer has been solubilized, addition of surfactant will result in a change in the mixed micelle composition as they become richer in surfactant in equilibrium with pure surfactant micelles. Both pure and mixed micelles are also in equilibrium with monomeric surfactant at the surfactant cmc for the system, i.e., that pertaining to the ionic strength and temperature. The values of R_e as defined by equation 6.1 will depend on the cmc. For a surfactant with a very low cmc, R_e will be approximately equal to the molar ratio of total surfactant to phospholipid, since [S] $\gg$ [S]$_{monomer}$.

The process of bilayer disruption can be followed by the release of an encapsulated marker from phospholipid vesicles as a function of surfactant concentration. Carboxyfluorescein is a very convenient marker molecule; it is highly fluorescent in dilute solution, but the fluorescence is quenched at higher concentration. When carboxyfluorescein is encapsulated in multilamellar vesicles at high concentration and the free carboxyfluorescein is removed by gel filtration, a population of vesicles with low (quenched) fluorescence can be obtained. The addition of surfactants to the vesicles results in an increased fluorescence as the bilayers of the multilamellar vesicles are solubilized and

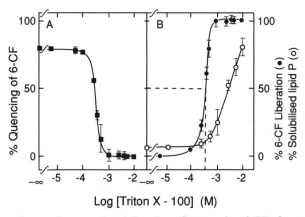

Log [Triton X - 100] (M)

Fig. 6.2. The release of encapsulated 6-carboxyfluorescein (6-CF) from phospholipid multilamellar liposomes by the nonionic surfactant Triton X-100. **A.** The decreases in fluorescence quenching on release of encapsulated 6-CF as a function of Triton X-100 concentration. **B.** % release of 6-CF and solubilization of phospholipid as a function of Triton X-100 concentration. The dotted line shows the procedure for defining R_{50}. From Ruiz et al. (1988), with permission of the publisher.

the encapsulated carboxyfluorescein is released. Figure 6.2 shows loss of quenching, the liberation of carboxyfluorescein, and the solubilization of lipid from egg yolk phosphatidylcholine liposomes as a function of Triton X-100 concentration. In these experiments, solubilization was measured by a filtration technique in which the intact vesicles were separated for solubilized lipid, which was assayed in the filtrate. It can be seen from Figure 6.2 that carboxyfluorescein liberation follows a steeper curve than solubilization, demonstrating that surfactant-induced leakage occurs at a lower surfactant concentration than solubilization. The molar concentrations of surfactant required to release 50% of the carboxyfluorescein and to solubilize 50% of the lipid can be determined from the plots and used to calculate the corresponding molar ratios of phospholipid to surfactant, R_{50} and S_{50}, respectively. Table 6.1 shows data for a number of commonly used surfactants. Studies of this type, using different encapsulated solutes and varying bilayer composition (Urbaneja et al., 1987; Alonso et al., 1987), show that these factors are important in assessing and determining the mechanism of solubilization; however, in broad terms, as the data in Table 6.1 show, the molar ratio of surfactant to phospholipid required to solubilize 50% of the bilayer increases with cmc of the surfactant. The more hydrophobic the surfactant, the lower will be its cmc and the greater will be its tendency to penetrate into the bilayer at low concentration. It is, however, important to note that the fine details of surfactant interactions with bilayers vary considerably with both the surfactant and the form of the bilayer. For example, in the case of sonicated unilamellar vesicles on interaction with Triton X-100, solubilization starts at a low surfactant concentration compared to that of larger vesicles, with the rapid formation of large multilamellar vesicles. At a surfactant to lipid molar ratio of 1:1, approximately a third of the lipid and most of the surfactant exist as soluble mixed micelles, while the remaining two-thirds of the lipid is in the form of multilamellar vesicles that are almost free of surfactant. These multilamellar vesicles are then solubilized as the surfactant to lipid ratio is further increased (Urbaneja et al., 1988).

TABLE 6.1. The Release of 6 Carboxy-Fluorescein and Solubilization of Lipid from Multilamellar Egg-Yolk Phosphatidylcholine Vesicles by Surfactants

Surfactant (cmc, mM)	R_{50}	S_{50}	$R_e^{(50\% \text{ sol.}) \text{a}}$
Triton X-100 (0.24)	2.9	0.60	1.7
Sodium n-dodecylsulphate (1.33)	0.83	0.40	2.5
Lauryldimethyl amino oxide (2.4)	1.3	0.29	3.4
Sodium cholate (3.0)	0.32	0.14	7.1
n-Octylglucoside (25)	0.09	0.05	20

From Ruiz et al. (1988).

[a] $R_e^{50\% \text{ sol}}$ was calculated from [S]/[PL], as the monomer concentration is not known at 50% solubilization.

6.1.2. The Mechanism of Surfactant-Induced Membrane Solubilization

The general mechanism of interaction of surfactants with a cell plasma membrane is shown in Figure 6.3. Here, as in the case of surfactant–phospholipid bilayer interactions, the details of the process occurring will differ from one membrane to another, so that the scheme depicted in Figure 6.3 is a generalized picture. For phospholipid bilayers, in the initial stages the surfactant partitions into the membrane bilayer, and as saturation is approached the cytoplasmic contents of the cell will leak out and the membrane will lyse. The soluble species formed on membrane disruption and solubilization will be largely mixed surfactant-lipid micelles and ternary complexes of lipid, protein (or glycoprotein), and surfactant. Increasing the surfactant concentration will, ideally, result in a mixture of surfactant-protein (glycoprotein) complexes, mixed micelles, and surfactant micelles. Whether or not the surfactant-protein complexes will be completely free of lipid will depend on the relative affinity constants of surfactant and lipid for the particular proteins present. The formation of completely lipid-free complexes is probably unlikely, especially for membrane proteins that have a high affinity for a particular lipid or a lipid requirement for their function.

As for the solubilization of phospholipid bilayers, the cmc is an important parameter in the lysis of cells. For example, Figure 6.4 shows the correlation of the release of 50% of cell protein from isolated fish gill epithelial cells with cmc for a range of surfactants (Partearroyo et al., 1991). In general terms, as for bilayer solubilization, the surfactant concentration required to release 50% of the cell protein increases with increase in the cmc. It is possible to distinguish

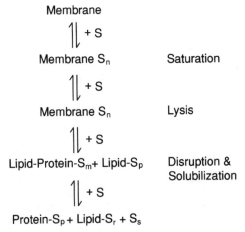

Fig. 6.3. Schematic representation of the sequence of events arising on exposure of a biomembrane to increasing amounts of surfactant (S). From Jones (1992), with permission of the Royal Society of Chemistry.

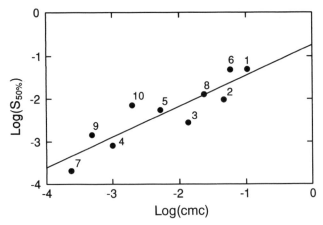

Fig. 6.4. Logarithm of the surfactant concentration required to solubilize 50% of the cell protein from isolated trout gill epithelial cells as a function of the logarithm of the critical micelle concentration. 1, sodium n-dodecylsulphate; 2, sodium n-decylsulphate; 3, sodium n-nonlysulphate; 4, sodium n-octyl sulphate, 5, n-dodecyltrimethylammonium bromide; 6, n-decyltrimethylammonium bromide; 7, Triton X-100; 8, n-octyl glucoside; 9, sodium deoxycholate; 10, sodium cholate. From Partearroyo et al. (1991), with permission of the publisher.

the effects of release of cell protein by lysis and membrane lipid solubilization by plotting the protein solubilization/phospholipid solubilization ratio as a function of surfactant concentration, as shown in Figure 6.5. For Triton X-100, sodium n-dodecylsulphate, and n-dodecyltrimethylammonium bromide, the protein to lipid ratios go through maxima at low surfactant concentrations,

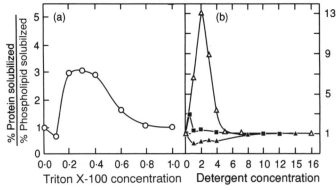

Fig. 6.5. Ratio of protein to phospholipid solubilized from isolated trout gill epithelial cells by surfactants in 30 min. **a.** ○, Triton X-100. **b.** ■, sodium n-dodecylsulphate; △, n-dodecyltrimethylammonium bromide; ▲, sodium cholate. From Partearroyo et al. (1991), with permission of the publisher.

showing that more protein is released than lipid solubilized. This is because the cell membranes become leaky to cytoplasmic protein as a consequence of partitioning of the monomeric surfactant into the membrane bilayer. As the surfactant concentration is increased, lipid is solubilized and the protein to lipid ratio tends to unity, consistent with the simultaneous solubilization of both cell protein and lipid. In contrast to these surfactants, the corresponding plot for sodium cholate passes through a minimum with surfactant concentration. This means that sodium cholate at low concentration can solubilize more membrane lipid than it can release protein, implying that lipid is removed from the membrane without it becoming leaky to cytoplasmic protein. This observation is in line with the ability of sodium cholate to form mixed micelles with lipid that contain a very high lipid to cholate ratios due to the stereochemistry of the cholate structure, which results in one side of the molecule being hydrophobic and the other hydrophilic (see Chaps. 1 and 3). Such effects are not restricted to gill epithelial cells; very similar solubilization behavior by sodium cholate has been found for human erythrocytes (Kirkpatrick et al., 1974) and human platelets (Shiao et al., 1989).

6.2. SURFACTANT-PROTEIN INTERACTIONS

As described above, the solubilization of cell membranes finally results in the formation of surfactant-protein and glycoprotein complexes in equilibrium with mixed surfactant–lipid and surfactant micelles. The analysis of surfactant-protein and glycoprotein complexes can easily be carried out by PAGE. This is routinely done after the samples have been treated with a reducing agent such as β-mercaptoethanol or dithiothreitol, which cleaves all the disulphide bonds and separates the proteins into their subunits. SDS is the preferred surfactant for PAGE. It is a powerful denaturant, so that all the tertiary structures are lost and the sample consists of SDS saturated (reduced) polypeptides, which bind SDS on a weight basis, generally of the order of 1.4–2.5 g SDS per g of protein. Under these conditions, the original electrostatic charge on the proteins is irrelevant, as the charge on the complexes in determined by the bound SDS. It is generally believed that for proteins the surfactant binds approximately uniformly along the polypeptide chain at a density of about one surfactant molecule per two amino acid residues, so that the charge per unit length is constant. When placed in an electrical field, the electrical force (charge × field strength) is proportional to the polypeptide chain length. The complexes will, however, be subjected to a frictional or viscous resistance, but this also increases with the polypeptide chain length. Thus the movement of such surfactant-saturated protein complexes in free solution when subject to an electric field is independent of polypeptide chain length. However, in a polyacrylamide gel the sieving action of the pores in the gel results in smaller complexes moving faster than the larger complexes, so that a separation in terms of molecular weight (mass) occurs. The separation of protein bands on a gel can be made visible by staining, usually with Coomassie blue stain. The value of SDS-PAGE is for the most part due to this molecular weight separation, which enables the polypeptide composition of the sample to be obtained and estimates of the molecular weights to be made with reference to standards.

The application of SDS-PAGE to glycoproteins is not as straightforward, because the oligosaccharide chains do not bind surfactant uniformly, so that glycoproteins are not separated in direct accordance with their chain length (molecular weight), making the interpretation of glycoprotein bands more difficult. Glycoproteins can be visualized on gels by staining either the protein (with Coomassie blue) or the oligosaccharide (with Periodic Shiff's reagent, PAS stain) or both, which helps in the interpretation of the results.

The formation of protein-surfactant complexes is of considerable interest, not only as a background to the mechanisms that lead to complexes in SDS-PAGE, but also as a means of looking at protein unfolding patterns. There is evidence for some globular proteins, such as ribonuclease A and lysozyme, that the surfactant-induced unfolding pathway resembles that of thermal unfolding, so that the study of surfactant-induced unfolding may give significant clues to unfolding processes in general and hence to protein stability (see Section 6.6).

6.2.1. Denaturing and Nondenaturing Surfactants

Sodium n-dodecylsulphate is a very potent protein denaturant; in general it will unfold most proteins at concentrations below its cmc (~ 8 mM in pure water). In contrast, denaturants such as urea or guanidinium chloride, which denature proteins as a consequence of the changes they bring about in the structure of water, cause denaturation only at very high concentration: 6–8 M for urea and 4–6 M for guanidinium chloride. SDS is thus 500–1,000 times more effective as a denaturant. There are however a few proteins that resist denaturation by SDS under some conditions; these include papain and pepsin (Nelson, 1971), glucose oxidase near neutrality (pH 6.0) (Jones et al., 1982b), and bacterial (*Micrococcus luteus*) catalase (Jones et al., 1982a), while *Aspergillus niger* catalase near neutrality (pH 6.4) is activated by SDS but not in acid (pH 3.2) or alkaline solution (pH 10.0) (Jones et al., 1987). Many synthetic ionic surfactants behave like SDS and can be classified as denaturing surfactants; the n-alkylsulphates, n-alkylsulphonates, n-alkyltrimethylammonium halides, and n-alkylpyridinum halides all act as denaturants to varying degrees. In contrast, the "natural" anionic surfactants—sodium cholates, and deoxycholates—and nonionic surfactants do not denature proteins. Although the cholates—and, for example, Triton X-100 and n-octylglucoside—will bind to the tertiary structure of proteins, they will not unfold them, so that function is retained in the case of enzymes, membrane transporters, and receptors (see Chapter 8). There are also cases of activation by nondenaturing surfactants; for example, Triton X-100 activates glucose-6-phosphatase (Beyhl, 1986) and sodium deoxycholate activates phospholipase (El-Sayert and Roberts, 1985).

With the classification of surfactants as denaturing or nondenaturing, the question arises as to origin of the distinction between the two types of behavior. Since all surfactants have an amphiphathic structure, why do ionic surfactants—excluding the natural ones—denature most proteins while nonionics do not? The answer clearly must lie in the nature of the headgroup interactions, at least in the initial stages of the interaction.

6.2.2. The Characterization and Mechanism of Surfactant Binding and Denaturation of Proteins

The overall mechanism of surfactant-induced denaturation of globular proteins is shown schematically in Figure 6.6 as it relates to ionic denaturing surfactant. The initial step involves the binding of the surfactant headgroup to ionic sites on the surface of the protein. Anionic surfactants will bind to cationic sites (lysyl, histidyl, and arginyl residues), while cationic surfactants will bind to anionic sites (glutamyl and aspartyl residues). While headgroup interactions with charged sites on the protein surface are electrostatic, the surfactants' alkyl chains interact hydrophobically with hydrophobic patches on the protein. The evidence for this type of initial interaction comes from binding isotherms which, for many globular proteins, show an initial plateau corresponding to saturation of the ionic residues (see below), together with studies on chemically modified proteins in which the effects of blocking the ionic sites and changing the alkyl chain length of the surfactant on the binding characteristics have been analyzed (Jones and Manley, 1980). The initial binding most probably occurs with little change in the tertiary structure of the protein and no change in the secondary structure. The protein-surfactant complexes will have more hydrophobic surface than the protein from which they are formed and at this stage may become insoluble and precipitate. However, increase in the surfactant concentration will result in further binding and the protein will denature. This unfolding will expose the hydrophobic residues previously buried in the tertiary structure, and surfactant will bind hydrophobically to these newly exposed sites. In general,

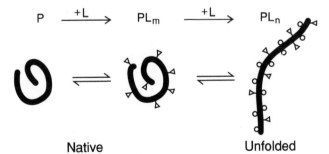

Fig. 6.6. A schematic representation of the binding of surfactant ligands (L) to the native state of a protein (P) and subsequent unfolding process. From Jones and Brass (1991), with permission of the Royal Society of Chemistry.

as the monomeric surfactant concentration approaches the cmc, the protein will become saturated with surfactant. This further binding almost always results in complexes that are soluble so that the initial precipitate formed dissolves.

Figure 6.7 shows typical binding isotherms for hen egg lysozyme on binding sodium n-dodecylsulphate (SDS) in acid solution, pH 3.2, 25°C. A binding isotherm in this field is conveniently displayed as the average number of surfactant molecules (or ions) bound per protein molecule ($\bar{\nu}$) as a function of the logarithm of the free surfactant concentration in equilibrium with the protein-surfactant complexes. Lysozyme (molecular mass 14,306) has a single polypeptide chain with an N-terminal lysine residue, four disulphide bonds, and 18 cationic residues (11 arginyl, 6 lysyl and 1 histidyl). At very low free SDS concentrations the binding isotherms are very steep, but reach a more gently rising plateau as the ionic interactions with the cationic residues saturate. As the cmc is approached, the binding isotherms rise steeply, characteristic of positive cooperativity. The effect of ionic strength on the binding isotherms is interesting in that, for the initial ionic interaction, increasing the ionic strength weakens the electrostatic interactions between protein and surfactant, so that the binding isotherm is shifted to higher free surfactant concentrations, while it strengthens the hydrophobic interactions, shifting the isotherm at high binding levels to low free surfactant concentration. In this sense, the basis for the division of ionic surfactant binding to proteins into specific ionic (polar) and hydrophobic (apolar) interactions is well founded, although in the intermediate region both types of binding may occur together.

Specific ionic binding will of course be significantly affected by changes in

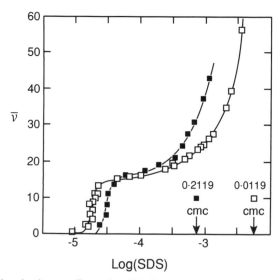

Fig. 6.7. Binding isotherms ($\bar{\nu}$ vs. log [SDS]) for the binding of sodium n-dodecylsulphate to lysozyme in aqueous solution at 25°C, pH 3.2; $\square$, ionic strength 0.0119 M; $\blacksquare$, ionic strength 0.2119 M. Adapted from Jones et al. (1984).

pH through changes in the state of ionization of the amino acid side chains. In acid solution, the cationic sites (lysyl pKa ~ 10, histidyl pKa ~ 6, arginyl pKa ~ 12) will be fully protonated while glutamyl and aspartyl side chains, which will interact repulsively toward anionic surfactants, will be partially protonated so that acidic conditions are very favorable for the binding of anionic surfactants. In alkaline solution the glutamyl and aspartyl residues will be fully ionized, and as the pH is increased the cationic sites will partially lose their positive charge, so that the protein will progressively lose affinity for specific ionic binding of anionic surfactant. It may still, however, bind hydrophobically, so that we may expect significant changes in the nature of binding isotherms with pH; thus for anionic surfactants at low pH, isotherms may display both specific polar and apolar binding characteristics, while at high pH only apolar binding will occur.

The binding of nonionic surfactants can only involve relatively weak head-group interactions, such as hydrogen bonding, so that the predominant inter-actions will be hydrophobic, involving the alkyl tails. These are generally insufficiently strong to bring about protein unfolding. In fact, in the case of globular proteins there is considerable disagreement as to whether nonionic surfactants such as n-octylglucoside bind to globular proteins at all (Cordoba et al., 1988; Lundahl et al., 1990), although there is no doubt that they bind hydrophobically to membrane proteins.

6.3. TECHNIQUES IN THE STUDY OF SURFACTANT-PROTEIN INTERACTION

There is a wide range of experimental techniques which can be used in the study of protein-surfactant interactions, as summarized in Table 6.2. The most important questions to be answered concern (a) the extent of surfactant binding, (b) the conformational changes induced by binding, and (c) the structure of the complexes that are formed. Of the methods for the measurement of binding, quantitative equilibrium dialysis is the most straightforward and thermody-namically sound. This involves placing a known volume of the protein solution of known concentration in a dialysis membrane bag and equilibrating it against a known volume of surfactant solution. The free concentration of surfactant in equilibrium with the complexes inside the bag is assayed by taking aliquots from outside the bag, since the free concentration is the same on both sides of the dialysis membrane. From the known initial amount of surfactant in the system and the final amount free, the amount bound to the protein can be calculated. The following experimental considerations are important in this technique: (1) the dialysis membrane tubing must be impermeable to the protein and its complexes but freely permeable to the surfactant; (2) the ionic strength must be sufficiently high to eliminate any Donnan effect, i.e., the development of an unequal concentration of surfactant on either side of the membrane to compensate for the charge on the protein; (3) the system must be in equilibrium

TABLE 6.2. Techniques Used in the Study of Surfactant–Globular Protein Interaction

Technique	Information Obtained
Quantitative equilibrium dialysis	Binding isotherms, Gibbs energy of ligand binding[a]
Molecular sieve chromatography	Binding levels[b]
Titrimetry	Proton binding in relation to surfactant binding[c]
Calorimetry (microcalorimetry and titration calorimetry)	Enthalpy of surfactant binding and protein unfolding[d]
Polyacrylamide gel electrophoresis	Detection of specific complexes[e]
Ultracentrifugation (sedimentation rate and equilibrium)	Sedimentation coefficients of protein–surfactant complexes, subunit dissociation and molecular weights[f]
Viscometry	Hydrodynamic volume and shape factors, protein unfolding[g]
Static and dynamic light scattering	Molecular weights, diffusion coefficients-complex dimensions[h]
Ultraviolet difference spectroscopy	Surfactant-induced conformational changes[i]
Neutron scattering	Structure of surfactant-protein complexes[j]
Enzyme kinetics	Surfactant-induced enzyme denaturation or activation[k]

Data from: [a]Jones and Manley (1979), Jones et al. (1984); [b]Lundahl et al. (1990); [c]Jones and Manley (1981, 1982); [d]Jones and Manley (1980), Finn et al. (1984), Kale et al. (1978); [e]Reboiras and Jones (1982); [f]Jones et al. (1982a, 1982b); [g]Tipping et al. (1974), Jones et al. (1975); [h]Jones and Midgley (1984), Makino et al. (1986); [i]Jones et al. (1973); [j]Ibel et al. (1990), Chen and Teixeira (1986); [k]Jones et al. (1987), Beyhl (1986).

at the chosen temperature; this may require at least 96 hours for a typical surfactant such as sodium n-dodecylsulphate. If the equilibrium dialysis experiment is carried out over a range of surfactant concentration, the resulting binding isotherm will give a considerable amount of useful information about the binding behavior, including the Gibbs energies of surfactant binding (see below). The binding of surfactant results in complexes which, if the surfactant bound was fully ionized, would have at high binding levels a high surface charge density; however, counterions binding will reduce this. Protons will always be one of the counterions in the case of complexes formed between proteins and anionic surfactants. Proton binding can be determined from titration curves of the protein in the presence of surfactant; such data are relatively difficult to interpret and require the surfactant binding isotherm over a range of pH, but in favorable circumstances the Gibbs energies of binding both protons and surfactant to the protein can be evaluated (Jones and Manley, 1982). To complete a thermodynamic analysis, the enthalpies of interaction can be measured by titration or microcalorimetry. A very sensitive instrument is required because the energies involved in binding monomeric surfactant below the cmc to a dilute protein solution ($<1\%$ w/v) are generally only of the order of 10–50 mJ.

The dissociation of protein subunits by surfactant will result in the formation of surfactant-subunit complexes. The dissociation can be followed by both ultracentrifugation sedimentation rate measurements, using absorption or schlieren optics, or by electrophoresis (PAGE). Conformational changes induced by surfactant binding can be followed by viscometry. The intrinsic viscosity ($[\eta]$) of a globular protein in its native state depends on its degree of hydration and asymmetry according to the Simha equation

$$[\eta] = \nu(\bar{v}_2 + \delta_1 v_1^0) \qquad (6.2)$$

where ν is the Simha factor, which for a rigid ellipsoid increases with its axial ratio (a/b) (for a sphere a/b = 1 and $\nu = 2.5$), $\bar{v}_2$ and v_1^0 are the partial specific and specific volumes of the protein and solvent, respectively, and δ_1 is the hydration parameter (grams of water per gram of dry proteins). For a roughly spherical (rigid) globular protein ($\bar{v}_2 \sim 0.7$ cm^3 g^{-1}) in aqueous media ($v_1 \sim$ 1 cm^3 g^{-1}), δ_1 is approximately 0.2 g g^{-1}; hence $[\eta] \sim 2.3$ cm^3 g^{-1}. On unfolding the structure becomes more flexible; the intrinsic viscosity of a flexible protein is given by the Flory-Fox equation

$$[\eta] = \Phi \frac{(\sqrt{r^2})^3}{M} \qquad (6.3)$$

in which Φ is a universal constant, $\sqrt{r^2}$ is the root-mean-square end-to-end distance of the polypeptide chain, and M the molecular mass. The root-mean-square end-to-end distance is molecular weight-dependent and proportional to M raised to a greater power than 1; hence $[\eta]$ for the unfolded protein is molecular weight-dependent, in contrast to the native state (equation 6.2). Surfactant-induced unfolding will thus result in significant increases in the intrinsic viscosity of the protein.

Of the other techniques to follow conformational changes, dynamic light scattering gives a diffusion coefficient that will decrease with the size of the structure, due to an increase in the frictional coefficient, while ultraviolet difference spectroscopy can be used to monitor changes in the environment of specific chromophores that absorb at ~ 280 nm, such as the aromatic amino acid residues phenylanaline, tyrosine, and tryptophan, arising from the surfactant binding. In the case of enzymes, the enzymic activity is often a convenient measure of unfolding, although it is also important to bear in mind that specific binding at the active site may result in enzyme inhibition before any appreciable conformational change takes place.

The determination of binding and conformational changes leaves the question of the detailed structure of complexes unanswered. At present there is no absolute method for structure determination of protein-surfactant complexes apart from X-ray diffraction, which has only been applied to lysozyme with three bound SDS molecules (Yonath et al., 1977). X-ray diffraction requires a crystal; in the case of lysozyme, cross-linked triclinic crystals of the protein were soaked in 1.1 M SDS and then transferred to water or a lower concentration (0.35 M) of SDS to allow the protein to refold. It was necessary to use cross-linked crystals to prevent them from dissolving when exposed to a high SDS concentration. The resulting denatured-renatured crystals were found to have three SDS molecules within a structure that was similar but not identical to that of native lysozyme. Neutron scattering has been applied in a few cases and is producing interesting results (see below), but it should be noted that it is a model-dependent technique.

6.4. THERMODYNAMICS OF SURFACTANT-PROTEIN INTERACTIONS

The formation of protein-surfactant complexes is an example of multiple equilibria between monomeric surfactant and a range of complexes having different numbers of bound surfactant molecules. These equilibria can be represented by a set of equations in which P is the protein and S the surfactant as follows:

$$P + S \rightleftharpoons PS_1$$

$$PS_1 + S \rightleftharpoons PS_2$$

$$PS_2 + S \rightleftharpoons PS_3$$

or in general

$$PS_{i-1} + S \rightleftharpoons PS_i \tag{6.4}$$

The equilibrium constants assuming ideality (activity coefficient of unity) for the first three steps, for example, are given by

$$K_1 = \frac{[PS_1]}{[P][S]}; \quad K_2 = \frac{[PS_2]}{[PS_1][S]}; \quad K_3 = \frac{[PS_3]}{[PS_2][S]} \tag{6.5}$$

Eliminating $[PS_1]$ and $[PS_2]$ by substitution of $[PS_1]$ in K_2 and $[PS_2]$ in K_3 gives

$$K_1 K_2 K_3 = \frac{[PS_3]}{[P][S]^3} \tag{6.6}$$

Hence, for a total of n steps

$$K_1 K_2 K_3 \ldots K_n = \frac{[PS_n]}{[P][S]^n} \tag{6.7}$$

If all the equilibrium constants were identical for the n equilibria ($K_1 = K_2 = K_3 \ldots = K_n = K$), then

$$K^n = \frac{[PS_n]}{[P][S]^n} \qquad (6.8)$$

The average number of surfactant molecules bound per protein molecule $\bar{\nu}$ would then be given by the concentration of bound surfactant divided by the total concentration of protein.

$$\bar{\nu} = \frac{\sum\limits_{i=1}^{n} i[PS_i]}{[P] + \sum\limits_{i=1}^{n} [PS_i]} \qquad (6.9)$$

Hence

$$\bar{\nu} = \frac{[P]\{[K_1][S] + 2K_1K_2[S]^2 \text{------} n(K_1K_2\text{---}K_n)[S]^n\}}{[P] + [P]\{K_1[S] + K_1K_2[S]^2 \text{----}(K_1K_2\text{---}K_n)[S]^n\}} \qquad (6.10)$$

While equation 6.10 is a rigorous equation for $\bar{\nu}$ as a function of surfactant concentration without values of the individual binding constants, it is of limited practical use. If, however, equilibrium constants were all equal, then from equation 6.8 it would follow that

$$\bar{\nu} = \frac{n[PS_n]}{[P] + [PS_n]} = \frac{n(K[S])^n}{1 + (K[S])^n} \qquad (6.11)$$

In general, since the equilibrium constants will not be equal, equation 6.11 would not be expected to hold but to preserve the simple form of the equation Hill introduced, a cooperativity coefficient n_H to replace n, giving (Hill, 1914)

$$\bar{\nu} = \frac{n(K[S])^{n_H}}{1 + (K[S])^{n_H}} \qquad (6.12)$$

For a positively cooperative process (i.e., binding of ligands enhances further binding) $n_H > 1$, while for a negatively cooperative process (i.e., binding of ligands inhibits further binding) $n_H < 1$.

An alternative approach to binding much used by biochemists is based on the Scatchard equation (Scatchard, 1949). If a protein has n independent and identical binding sites with intrinsic binding constants K and a fraction Θ of these are occupied at a given surfactant concentration [S], then a simple kinetic argument in which the rate of binding proportional to [S] times the fraction, of vacant sites $(1 - \Theta)$, is equated to the rate of dissociation from the occupied sites, proportional to Θ, gives

$$\Theta = K[S](1 - \Theta) \qquad (6.13)$$

Since $\Theta = \bar{\nu}/n$; with a little manipulation it follows that

$$\frac{\bar{\nu}}{n} = K[S]\left(1 - \frac{\bar{\nu}}{n}\right) \qquad (6.14)$$

hence

$$\bar{\nu} = \frac{n\,K[S]}{1 + K[S]} \qquad (6.15)$$

Equation 6.15 is identical to the Hill equation (6.12) when the Hill coefficient is unity. It follows from the Scatchard equation that

$$\frac{\bar{\nu}}{[S]} = K(n - \bar{\nu}) \qquad (6.16)$$

so that the plot of $\bar{\nu}/[S]$ is a linear function of $\bar{\nu}$ with slope K and intercept on the abscissa of n.

It is informative to see how the shape of binding isotherms and their corresponding Scatchard plots based on equation 6.16 depend on cooperativity. Figure 6.8 shows a series of binding isotherms generated from the Hill equation (6.12) for a hypothetical molecule (e.g., protein) with 50 binding sites with intrinsic binding constant 10^4 for various Hill coefficients from 0.5 (negative cooperativity) to 7.5 (highly positive cooperativity) (Jones and Brass, 1990). The curves are sigmoidal, increasing in steepness with increasing n_H. There is no way of distinguishing between negative and positive cooperativity from inspection, since the curves are qualitatively similar. In contrast, the Scatchard plots for the same systems show dramatically changing characteristics (Fig. 6.9) that are diagnostic of the type of cooperativity (Schwarz, 1976). For Hill coefficients less than unity, the Scatchard plots have negative slopes decreasing with increasing $\bar{\nu}$ and intercept the abscissa at n(50); for $n_H = 1$ the plot is linear as required by equation 6.16, whereas for $n_H > 1$ the plots pass through maxima, the maximum point becoming asymmetrically displaced to higher values of $\bar{\nu}$ with increasing n_H. The misuse of the Scatchard plot for $n_H < 1$ has been the subject of much discussion (Klotz, 1982; Klotz and Hunston, 1984; Feldman, 1983). The problem relates primarily to extrapolation of the roughly linear part of the plot of $n_H < 1$ to the abscissa to obtain an intercept, which is taken as the total number of binding sites. For example, in Figure 6.9 for $n_H = 0.5$ the steep part of the plot could be extrapolated to give n ~ 12–13, but such an extrapolation has no significance in this context.

When the Scatchard analysis is applied to protein-surfactant interactions,

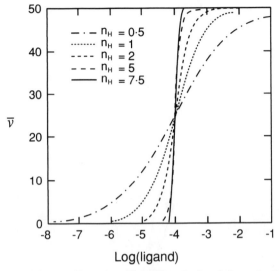

Fig. 6.8. Binding isotherms ($\bar{v}$ vs. log [ligand]) calculated from the Hill equation for a protein with 50 binding sites (intrinsic binding constant 10^4) for a range of Hill coefficients from 0.5 to 7.5. From Jones and Brass (1991), with permission of the Royal Society of Chemistry.

examples of both negative and positive cooperativity are found for some systems. Figures 6.10 and 6.11 show Scatchard plots for the binding of SDS to bovine catalase (mol. mass 245,000) in acid solutions. At pH 3.2 and 4.3, curves diagnostic of negative cooperativity are obtained that can be extrapolated to give values of n of 343 ± 6 and 333 ± 13, respectively, which are close to the number of cationic amino acid residues in the catalase molecule of 331

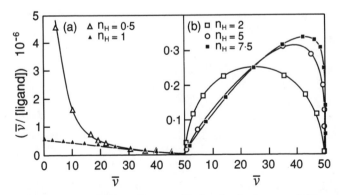

Fig. 6.9. Scatchard plots ($\bar{v}$/[ligand]$_{\text{free}}$ vs. $\bar{v}$) for the isotherms of Figure 6.8 for a protein with 50 binding sites (intrinsic binding constant 10^4) for a range of Hill coefficients from 0.5 to 7.5. From Jones and Brass (1991), with permission of the Royal Society of Chemistry.

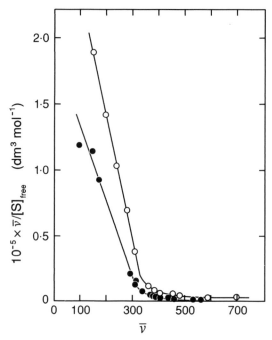

Fig. 6.10. Scatchard plots for sodium n-dodecylsulphate (SDS) on binding to bovine catalases at 25°C. $\bigcirc$, pH 3.2; $\bullet$, pH 4.3. Adapted from Jones and Manley (1982).

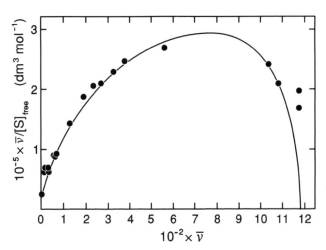

Fig. 6.11. Scatchard plots for sodium n-dodecylsulphate (SDS) on binding to bovine catalase at pH 6.4, 25°C. The solid line was fitted using the Hill equation for a total of 1,190 binding sites (1.4 g SDS per g catalase), giving an intrinsic binding constant 479 ± 6 dm^{-3} mol^{-1} and $n_H = 2.62 \pm 0.07$. Adapted from Jones and Manley (1982).

(112 lysyl, 86 histidyl, and 133 arginyl). The flatter parts of the curves correspond to further binding, largely of a hydrophobic type. At pH 6.4, a typical positively cooperative Scatchard plot is found corresponding to a Hill coefficient of 2.61 ± 0.07. Table 6.3 shows data for other proteins obtained by Scatchard analysis where, like catalase, extrapolation gives values of the total number of specific binding sites very close to the number of cationic amino acid residues in the protein. The shape of the linear part of the Scatchard plots gives values for the intrinsic binding constant, and hence Gibbs energies of binding per mole of SDS ($\Delta G_{\bar{v}}$). Also shown in Table 6.3 are the corresponding enthalpies of binding per mole of SDS ($\Delta H_{\bar{v}}$) measured by microcalorimetry, which combined with $\Delta G_{\bar{v}}$ give $T\Delta S_{\bar{v}}$. It can be seen that the enthalpies of binding are in general exothermic, but small relative to $T\Delta S_{\bar{v}}$. The large increases of entropy on binding are characteristic of a substantial hydrophobic contribution to the binding process arising from the disordering of water molecules, concomitant with the partial removal of the alkyl chains of the surfactant from the aqueous environment. Thus initial binding of surfactants to proteins requires, not only the ionic interaction of headgroups with cationic sites, but also binding of the alkyl chains to hydrophobic regions of the protein in the vicinity of the cationic sites. Confirmation of this comes from the observations that chemical modification of the cationic sites—e.g., in the case of lysyl residues by acetylation—shifts the Scatchard plots to give lower values of n and that reducing the alkyl chain length weakens binding (Jones and Manley, 1980). Despite these observations, any Scatchard analysis should be treated with a degree of caution; it is not entirely clear why such a good correspondence between n and the number of cationic sites is obtained, since the binding sites must only approximate to independence and are certainly not chemically identical.

TABLE 6.3. Scatchard Analysis of the Binding of Sodium n-Dodecylsulphate to Some Globular Proteins

Protein (Molecular Weight), pH	No. of Cationic Residues	n	K (dm³ mol⁻¹)	$\Delta G_{\bar{v}}$ (kJ mol⁻¹) (-RT ln K)	$\Delta H_{\bar{v}}$ (kJ mol⁻¹)	$T\Delta S_{\bar{v}}$ (kJ mol⁻¹)
Ribonuclease A (13,682), 7	18	19	7.70×10^4	−27.9	−1.27	26.6
Lysozyme (14,306), 3.2	18	18	3.93×10^4	−26.2	−8.66	17.5
Ovalbumin (44,000), 7.0	42	37	1.77×10^5	−30.0	0	30.0
Glucose oxidase (147,000), 3.7	120	132	0.51×10^4	−21.2	−3.29	17.9
Bovine catalase (245,000), 3.2	331	343	9.53×10^4	−28.4	−8.36	20.0

From Jones (1988).

A more rigorous analysis of binding isotherms can be made by the application of the binding potential concept proposed by Wyman (1965). Wyman recognized that there should be a function, which he called the binding potential π (P, T, μ_i, μ_j----), relating the extent of binding (ν) to the chemical potential of any ligand (μ) such that

$$\nu = \left(\frac{\partial \pi}{\partial \mu} \right)_{P, T, \mu_i, \mu_j---} \tag{6.17}$$

If the chemical potential of a surfactant ligand (μ_s) is given by the ideal solution expression

$$\mu_s = \mu_s^0 + RT \ln [S] \tag{6.18}$$

where μ_s^0 is the standard chemical potential, here at unit molarity, then it follows from equations 6.17 and 6.18 that

$$\nu = \frac{1}{RT} \left(\frac{\partial \pi}{\partial \ln [S]} \right)_{P, T, \mu_i, \mu_j---} \tag{6.19}$$

Hence the binding potential will be given by

$$\pi = 2.303 \ RT \int_0^\nu \bar{\nu} \ d \log [S] \tag{6.20}$$

It follows from equation (6.20) that the binding potential can be calculated by integrating under the binding isotherm ($\bar{\nu}$ vs. log [S]) from $\bar{\nu} = 0$ to any desired value of $\bar{\nu}$. If $\bar{\nu}$ is expressed by equation 6.11 then by differentiation it can be shown that

$$\frac{d \ln [S]}{d \bar{\nu}} = \frac{1}{K^n [S]^n (n - \bar{\nu})^2} \tag{6.21}$$

and, by substitution of equation 6.21 into 6.20 followed by integration,

$$\pi = 2.303 RT \log \left(\frac{n}{n - \bar{\nu}} \right) \tag{6.22}$$

which on manipulation using equation 6.11 gives

$$\pi = 2.303 RT \log (1 + K^n [S]^n) \tag{6.23}$$

If it is assumed that for any given free surfactant concentration for an extent of binding $\bar{\nu}$ corresponding to a complex $PS_{\bar{\nu}}$ then

$$\pi = 2.303RT \log (1 + K_{app}[S]^{\bar{\nu}}) \qquad (6.24)$$

Since π can be evaluated by integration under the binding isotherm at any desired value of $\bar{\nu}$ corresponding to a particular free surfactant concentration, it is possible to determine a range of apparent binding constants, K_{app}, as binding proceeds. The Gibbs energy of binding per ligand bound can then be determined from

$$\Delta G_{\nu} = -\frac{RT}{\bar{\nu}} \ln K_{app} \qquad (6.25)$$

Figure 6.12a shows the results of application of this approach to the binding isotherm for the hypothetical molecule with 50 binding sites depicted in Figure 6.8 for a range of cooperativity coefficients. For a Hill coefficient of unity and $K_{app} = 10^4$, $\Delta G_{\bar{\nu}}$ should be independent of $\bar{\nu}$ and have a constant value of -22 KJmol^{-1}. For negative cooperativity ($n_H < 1$), $\Delta G_{\bar{\nu}}$ should become less negative with increasing $\bar{\nu}$. It can be seen, however, that all the isotherms lead to plots of $\Delta G_{\bar{\nu}}$ which become less negative with increasing $\bar{\nu}$. The reason for this is the neglect of the statistical contributions to $\Delta G_{\bar{\nu}}$. For example, when binding the first ligand to a protein with 50 sites, there are 50 possible locations; two ligands can be placed in $50 \times (49/2) = 1,225$ different ways, if the sites are indistinguishable. In general, the number of arrangements $\Omega_{n,i}$ of i ligands on n indistinguishable binding sites is given by

$$\Omega_{n,i} = \frac{n!}{(n-i)!i!} \qquad (6.26)$$

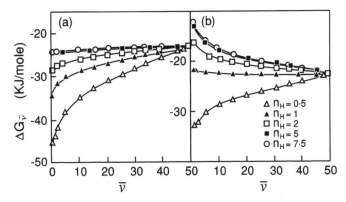

Fig. 6.12. Gibbs energies of binding per ligand bound ($\Delta G_{\bar{\nu}}$ versus $\bar{\nu}$) calculated by the Wyman binding potential method from the isotherms of Figure 6.8 for a protein with 50 binding sites (intrinsic binding constant 10^4) for a range of Hill coefficients from 0.5 to 5.0. **a.** Without statistical corrections. **b.** With statistical corrections. From Jones and Brass (1991), with permission of the Royal Society of Chemistry.

This gives an entropic contribution of R ln $\Omega_{n,i}$ and hence a Gibbs energy contribution per ligand bound given by

$$(\Delta G_{\bar{\nu}})_{\text{statistical}} = -\frac{RT}{i} \ln \Omega_{n,i} \qquad (6.27)$$

When the statistical contribution is added to $\Delta G_{\bar{\nu}}$ calculated from equation (6.27), the plots of $\Delta G_{\bar{\nu}}$ vs. ν follow the expected trends, becoming less negative for negative cooperativity ($n_H > 1$) (Fig. 6.12b). The curves converge to -22 kJ mol^{-1} ($K_{app} = 10^4$) when $\bar{\nu} = 50$, since the statistical effects decrease with $\bar{\nu}$ as the number of possible arrangements of the ligands on vacant sites decreases as the sites are filled.

The above discussion has been concerned with statistical effects in a model system. The situation is less clear in the case of a real system in which the protein undergoes surfactant-induced denaturation. In such cases, the number of binding sites is not constant but increases when the protein unfolds. If unfolding occurs cooperatively over a very narrow region of the binding isotherm, then the appropriate statistical corrections can be made. Considering lysozyme, for example, the binding isotherm (Fig. 6.7) shows that the 18 cationic sites largely saturate before cooperative hydrophobic binding occurs and the isotherm rises steeply again. The range of $\bar{\nu}$ over which unfolding occurs is relatively narrow, since denaturation is essentially a two-state process for lysozyme. Although we cannot define this range precisely, calorimetric measurements indicate that unfolding occurs close to the point of saturation of the specific binding sites. Thus the statistical corrections to the Gibbs energy per surfactant bound must be made in two parts; first, correction is made to the Gibbs energy of binding to the first 17–18 specific sites on the native protein, and second, to the remaining sites exposed on unfolding. The binding isotherm suggests that the number of binding sites at saturation is ~60 (this figure corresponds to the binding of 1.2 g of SDS per g of lysozyme and is consistent with saturation binding levels to nonreduced globular proteins). If it is assumed that surfactant remains bound to the specific cationic sites on and after unfolding, then the statistical corrections to the unfolded state must be applied to approximately 42 sites. Figure 6.13 shows the results of the application of the Wyman approach to SDS binding to lysozyme at two ionic strengths before and after correction for the statistical effects. The statistical corrections introduce a discontinuity in the plots associated with unfolding, from which it is possible to estimate the Gibbs energy of unfolding of the liganded protein. If we consider the binding of $\bar{\nu}$ ligands (L) to the native (N) and unfolded state (U) of the protein according to the equations

$$N + \bar{\nu}L \rightleftharpoons N(L)_{\bar{\nu}} \qquad \Delta G_b^N \qquad (6.28)$$

$$U + \bar{\nu}L \rightleftharpoons U(L)_{\bar{\nu}} \qquad \Delta G_b^u \qquad (6.29)$$

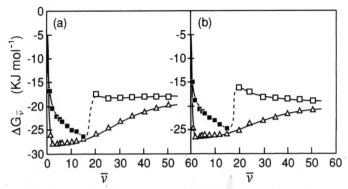

Fig. 6.13. Gibbs energy of binding per sodium n-dodecylsulphate ligand as a function of the number of SDS ligands bound to lysozyme in aqueous solution at 25°C, pH 3.2. Δ, calculated from the binding potential without statistical corrections; □, ■, calculated with statistical corrections for binding to the native (■) and unfolded (□) states. **a.** Ionic strength 0.0119 M. **b.** Ionic strength 0.2119 M. From Jones and Brass (1991), with permission of the Royal Society of Chemistry.

where ΔG_b^N and ΔG_b^u are the Gibbs energies of binding, then

$$\Delta G_b^N = \Delta G_b^{N,o} + RT \ln \frac{[N(L)_{\bar{\nu}}]}{[N][L]^{\bar{\nu}}} \tag{6.30}$$

$$\Delta G_b^U = \Delta G_b^{U,o} + RT \ln \frac{[U(L)_{\bar{\nu}}]}{[U][L]^{\bar{\nu}}} \tag{6.31}$$

where $\Delta G_b^{N,o}$ and $\Delta G_b^{U,o}$ are the standard Gibbs energies of binding. Then, subtracting equation 6.30 from equation 6.31 and rearranging, we obtain

$$\Delta G_b^U - \Delta G_b^N = \Delta G_b^{U,o} - \Delta G_b^{N,o} + RT \ln \frac{[N]}{[U]} + RT \ln \frac{[U(L)_{\bar{\nu}}]}{[N(L)_{\bar{\nu}}]} \tag{6.32}$$

At the midpoint of the transition from the native to the unfolded state, $\Delta G_b^U = \Delta G_b^N$. The other terms may also be replaced as follows:

$$\Delta G_b^{U,o} - \Delta G_b^{N,o} = \nu(\Delta G_{\bar{\nu}}^{U,o} - \Delta G_{\bar{\nu}}^{N,o}) = \bar{\nu}(\delta \Delta G_{\bar{\nu}}) \tag{6.33}$$

$$RT \ln \frac{[N]}{[U]} = \Delta G_U^o \tag{6.34}$$

$$RT \ln \frac{[U(L)_{\bar{\nu}}]}{[N(L)_{\bar{\nu}}]} = -\Delta G_{u/L}^o \tag{6.35}$$

when ΔG_U^o and $\Delta G_{U/L}^o$ are the standard Gibbs energies of unfolding of the unliganded and liganded protein, respectively. Hence from equation (6.32)

$$0 = \bar{\nu}(\delta\Delta G_{\bar{\nu}}) + \Delta G_U^o - \Delta G_{U/L}^o \qquad (6.36)$$

or

$$\bar{\nu}(\delta\Delta G_{\bar{\nu}}) = \Delta G_{U/L}^o - \Delta G_U^o \qquad (6.37)$$

Equation 6.37 shows that from the change in $\Delta G_{\bar{\nu}}$ at the discontinuity ($\delta\Delta G_{\bar{\nu}}$) in the plot of $\Delta G_{\bar{\nu}}$ vs. $\bar{\nu}$, the standard Gibbs energy of unfolding of the liganded protein can be deduced if the standard Gibbs energy of unliganded protein is known.

Table 6.4 shows results obtained by application of the above arguments to data for the binding of SDS to lysozyme (Fig. 6.13) covering a range of ionic strengths. The values of $\Delta G_{U/SDS}^o$ are very much larger than the values of ΔG_U^o, which shows that the initial binding of surfactant stabilizes the native state relative to the unliganded native molecule, but as binding proceeds the decrease in Gibbs energy resulting from unfolding and exposure of a large number of hydrophobic binding sites more than compensates for the energy required to unfold the liganded complex. This can be seen in Figure 6.14, where the Gibbs energy of complex formation ($\Delta G = \bar{\nu}\Delta G_{\bar{\nu}}$) is plotted as a function of $\bar{\nu}$. The "blip" in the plot shows the transition from the liganded native to the liganded unfolded state.

The application of the Wyman binding potential method to protein-surfactant interaction gives an interesting insight into the complex process of multiple equilibria as represented by equation 6.4. It can be applied to any binding isotherm to give a $\Delta G_{\bar{\nu}} - \bar{\nu}$ profile. The introduction of statistical corrections, however, requires additional information about the range of $\bar{\nu}$ over which unfolding occurs and the numbers of binding sites in the native and unfolded states; it also requires additional data about the unfolding process, which must

TABLE 6.4. Thermodynamic Parameters for the Lysozyme-SDS Interaction in Aqueous Solution (pH 3.2) at 25°C

Ionic strength (M)	$\bar{\nu}$ (transition pt.)	$\delta\Delta G_{\bar{\nu}}$ (kJ mol^{-1})[a]	$\nu(\delta\Delta G_{\bar{\nu}})$ (kJ mol^{-1})[a]	ΔG_U^o (kJ mol^{-1})[a]	$\Delta G_{U/SDS}^o$ (kJ mol^{-1})[a]
0.0119	17	10	170	46	216
0.0269	18	11.5	207	46	253
0.0554	17	10	107	46	216
0.1119	17	10.5	179	46	225
0.2119	18	10.5	189	46	235
Average					229 ± 16

[a]Data from Pfeil and Privalov (1976).

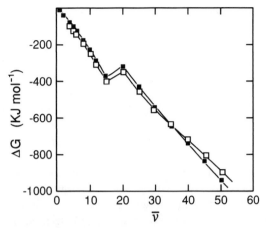

Fig. 6.14. Gibbs energy of formation of lysozyme-sodium-n-dodecylsulphate complexes as a function of the number of SDS ligands bound at 25°C, pH 3.2. □, ionic strength 0.0119 M; ■, ionic strength 0.2119 M. From Jones and Brass (1991), with permission of the Royal Society of Chemistry.

come from calorimetry, spectroscopy, or possibly transport measurements (viscometry). In this sense the corrections for the statistical effects are more problematical in that they are subject to the precise details of the surfactant-induced unfolding process.

6.5. THE STRUCTURE OF PROTEIN-SURFACTANT COMPLEXES

The structure of protein-surfactant complexes has to date proved very difficult to determine unambiguously, particularly for globular proteins with their disulphide linkages intact. The situation is a little clearer in the case of reduced proteins. The success of the SDS-PAGE technique, which depends on the existence of a uniform charge per unit length for the saturated SDS-polypeptide complexes, suggests that the complexes must have an extended conformation, with surfactant molecules bound approximately uniformly along their length. At the other extreme, when only a few molecules of surfactant are bound to an oxidized (disulphide bonds intact) protein, the protein structure will be very little different from that of the native state. In the case of lysozyme-SDS complexes with a few bound SDS molecules, this has been largely confirmed from the X-ray crystallographic structures (Yonath et al., 1977). To obtain crystallographic data on the lysozyme-SDS complexes, it was necessary to cross-link triclinic lysozyme crystals and to soak them in SDS (1.1 M) before transferring them to water or a lower concentration of SDS solution (0.35 M) to allow the protein to refold. Examination of the resulting "denaturant-renatured" crystals by X-ray diffraction revealed three SDS molecules in the renatured structure; however, the renatured structure was similar but not identical to native-state crystals. The agreement between the structure factors of the renatured crystals and the native cross-linked crystals was 17% (for renaturation in water) and 19% (for renaturation in 0.35 m SDS), and the minimum spacings in the X-ray pattern of renatured and native crystals were 2.9 Å and 1.1 Å, respectively. The need to cross-link the crystals detracts somewhat from the significance of the results, although there were only a few cross-links and these were highly flexible. The results clearly showed the location of the SDS molecules, with the surfactant headgroups forming a salt bridge with positively charged amino acid residues and the hydrocarbon chains making hydrophobic contact with the tertiary structure. Specifically, of the three SDS molecules bound in the renatured crystals, one formed a salt-bridge with the terminal lysine, with its alkyl chain penetrating deep into the hydrophobic core of the tertiary structure, while the other two were bound to the protein surface, one being shared between two lysozyme molecules in the renatured crystal. The SDS molecule that penetrated into the hydrophobic core could not be removed even after soaking in SDS-free water for an extended period. Lysozyme has two domains separated by a cleft into which the natural substrate (cell-wall polysaccharide) binds. During renaturation, the two domains fold separately, trapping the SDS molecule between them.

The structures of complexes lying between the two extremes, represented by oxidized globular proteins with a few bound surfactant molecules and reduced globular proteins saturated with surfactant, are much more difficult to study. A variety of models have been proposed and usefully summarized by Ibel et al. (1990) as follows:

1. A model based on a "micellar complex" in which the protein assembles the surfactant molecules to form a micelle of definite size.
2. A "rodlike-particle" model in which the polypeptide chain forms the core of a rod of 3.6 nm diameter with surfactant bound along its length. The rod is not totally rigid and has flexible regions between short rigid segments.
3. A "pearl-necklace model" in which the polypeptide chain forms the string of the necklace and the surfactant molecules form micelle-like clusters along the polypeptide chain, which passes through the micellar clusters in an α-helical conformation. In contrast to the rodlike-particle model, the model assumes the polypeptide chain is flexible.
4. An "α helix-random coil model" in which the surfactant binding enhances the α-helical content of the protein and disrupts the β structure.
5. A "flexible helix model" in which the surfactant molecules form a flexible cylindrical micelle and the polypeptide chain of the protein wraps around it and is stabilized by hydrogen bonding between the surfactant headgroup and the peptide-bond nitrogen atoms.

It is clear from a consideration of the diversity of the proposed models that our understanding of the detailed structure of surfactant-protein complexes is far from adequate, despite very many years of study. The pearl-necklace model and the flexible helix model could hardly be more different. Neutron scattering has been used in an attempt to get more direct information on the structure of complexes, although here again the results must be fitted to models so that the conclusions are not unambiguous. In the case of complexes formed between bovine serum albumin (BSA) and lithium n-dodecylsulphate, the small-angle neutron scattering data were interpreted in terms of the pearl necklace model, in which micelles of 1.8 nm radius (aggregation number 70 ± 20) were distributed along the single polypeptide chain of BSA (Chen and Teixeira, 1986). The correlations between the micelles are given by the structure factor

$$S(Q) = 1 + N_p \int_o^\infty 4\pi r^2 g(r) \frac{SinQr}{Qr} \, dr \qquad (6.38)$$

where N_p is the number density of micelles of radius R and $g(r)$ is the pair-correlation function, which is related to $N(r)$, the number of individual scatterers within a sphere of radius r,

$$g(r) = \frac{1}{4\pi r^2} \left(\frac{dN}{dr} \right) \qquad (6.39)$$

$N(r)$ can be related to the fractal dimension (D) of the protein-surfactant complex by the equation $N(r) = (r/R)^D$. For a freely diffusing micelle the fractal dimension would be 3, but in a complex the distribution of micelles is dictated by the topology of the polypeptide backbone, so that the fractal dimension is less than 3. At 1% by weight of lithium n-dodecylsulphate, D is 2.3 and decreases to 1.76 at 3%, consistent with a transition from a compact state to a more open random coil in which a string of constant-sized micelles are distributed along the hydrophobic patches of a denatured random coil, although the coil will be restrained by the 17 disulphide bonds in the BSA structure.

The pearl necklace model for the BSA-dodecylsulphate complexes is very different from the model used to interpret the neutron diffraction data for the complexes formed between the deuterated bifunctional enzyme N-5'-phosphoribosylanthranilate/indole-3-glycerol-phosphate synthase (PRA-IGP) and SDS (Ibel et al., 1990). This enzyme from *Escherichia coli* contains a single polypeptide chain of 452 amino acids (mol. wt. 49,484) with no disulphide bonds. The deuterated molecule binds 1.26 g SDS per g protein (216 SDS molecules/452 amino acid residues). Neutron scattering was investigated from the whole molecule complex (W) and two SDS-complexed fragments produced by gentle hydrolysis with trypsin, a large fragment (L) containing 289 residues and a small fragment (S) containing 163 residues. Figure 6.15 shows the pair-distance distribution functions (PDDFs) of volume elements of the three SDS-complexed structures at vanishing contrast between the buffer-medium and the protein-surfactant phase. The small complex (S) gives a single peak corresponding to a single globular structure and neutron scattering total dodecyl-

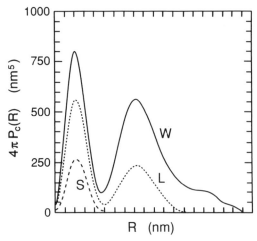

Fig. 6.15. Pair-distance distribution functions (PDDFs) of volume elements situated in *n*-dodecylsulphate phases of SDS-(N-5′-phosphoribosylanthranilate isomerase/indole-3-glycerol-phosphate synthase) complexes as observed by neutron scattering at vanishing contrast between the buffer and the protein-SDS phase. S, small fragment, molecular weight 17,478, 163 amino acid residues. L, large fragment, molecular weight 32,024, 289 amino acid residues. W, whole molecule, molecular weight 49,484, 452 amino acid residues. Adapted from Ibel et al. (1990).

chain volume (V_c) of 26.0 ± 1 nm^3 and 73 ± 3 $C_{12}H_{25}$ chains. The large complex (L) gives two peaks, which arise from two micelles associated with the C- and N-terminal ends of the polypeptide chain (L_C and L_N), which are well separated. The first peak (the self peak) is due to interferences of pairs of volume elements within the same micelle, and the second peak is due to interferences of pairs of volume elements in the micelles associated with L_C and L_N; for these, V_c is 35.6 ± 1.5 nm^3 (101 ± 4 SDS molecules) and 14.7 ± 0.8 nm^3 ($42 \pm$ SDS molecules), respectively. The whole molecule complex (W) gives two peaks and a shoulder; these are interpreted as the self peak (at low R), the central peak arising from interferences between pairs of volume elements situated in the core of a central micelle (W_M) and the core of a micelle associated with either the C-terminal (W_C) or N-terminal (W_N) end of the polypeptide chain and, finally, the shoulder (at large R) arising from interference between pairs of volume elements in W_C and W_N. The number of SDS molecules in the three micelles W_M, W_C and W_N are approximately 42, 101, and 73, respectively, giving the proposed structure shown in Figure 6.16 in which the polypeptide chain is wrapped around the three micelles to give what is described as a "protein-decorated micelle" structure. In the structure it is assumed that the two interconnecting polypeptide segments, which may bind a small number of SDS molecules, are highly flexible—as in the pearl necklace model—and the repulsive interaction between the micelles leads to an overall elongated conformation.

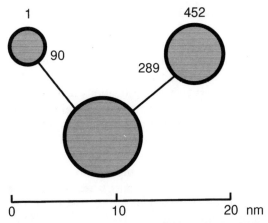

Fig. 6.16. "Protein-decorated micelle" model of the complex formed between 216 sodium *n*-dodecylsulphate molecules and the 452 amino acid residues of N-5'-phosphoribosylanthranilate isomerase/indole-3-glycerol-phosphate synthase. The complex consists of three spherical micelles (gray areas) mostly of SDS alkyl chains with hydrophilic shells (black areas) occupied by polypeptide chains and sulphate headgroups. Adapted from Ibel et al. (1990).

There is clearly a considerable difference between the pearl necklace and decorated micelle models, which may in part relate to the differences between the proteins, in particular the fact that BSA has a more restricted conformation because of disulphide linkages. However, the most important difference is that in the pearl necklace model the polypeptide chain is believed to pass through micelles of constant size, as opposed to around micelles of variable size in the decorated micelle model. However, it is significant that for the decorated micelles, the number of SDS molecules per amino acid residue is surprisingly uniform: 0.45 (S), 0.49 (L), and 0.48 (W).

A model approximating to a decorated micelle is suggested for the complexes formed on interaction of apocytochrome c and its fragments with SDS (Snel et al., 1991). Apocytochrome c is the precursor of mitochondrial cytochrome c; it is highly basic and interacts with negatively charged lipid. After synthesis in the cytoplasm, it is translocated across the outer mitochondrial membrane by a process coupled to covalent attachment of a heme group catalyzed by cytochrome c heme lyase. During insertion into the membrane, it undergoes a random coil to α-helix transition, which has been observed by circular dichroism. From studies in which apocytochrome c has been inserted into phospholipid vesicles, it was found that the amount of α-helix induced is close to 22% for dioleoylphosphatidylserine vesicles; interestingly, a very similar amount of α-helix induction occurs when apocytochrome c interacts with SDS micelles, suggesting that the topologies of the protein in the lipid and SDS micellar environments are similar. Photochemically induced dynamic nuclear polarization ^{1}H NMR spectroscopy of histidine, tryptophan, and tyrosine residues showed enhancement of signals in the presence of SDS micelles, while fluorescence spectroscopy showed quenching of tryptophan and tyrosine by micelles. These results suggest that the aromatic residues are localized in the interface of the SDS micelles, although the possible insertion of the α-helices into the micelle cannot be ruled out.

Despite the relatively extensive investigations that have been carried out on the structure of protein-surfactant complexes, definitive results have yet to be obtained. It is unfortunate that many of the investigations using modern sophisticated physical methods such as neutron diffraction have largely been done without reference to the exact state of binding as represented by the binding isotherm. As the extent of binding changes with surfactant concentration, so will the topology of the complexes, and only when studies are made with careful reference to the binding isotherms will we be able to obtain a clearer picture of the structure of the complexes.

6.6. MOLECULAR DYNAMICS OF PROTEIN-SURFACTANT COMPLEXES

A recent theoretical development in the study of protein-surfactant complexes is the application of molecular dynamics (McCammon and Harvey, 1987). This technique enables the energies of interaction between protein and surfactant to be calculated from first principles by solving the equations of motion for each atom in the complex, the unliganded protein, and the surfactant molecule, provided the potential functions describing the molecular interactions are known. From a knowledge of such potential functions, the force on each atom at some time t can be calculated, and then from Newton's equations of motion it is possible to calculate the acceleration of each atom, which on integration over a specified time interval (δt) gives the velocity and then the position of each atom at $t + \delta t$. These values are then used to repeat the calculation, and by such an interative procedure the conformation of minimum energy can be obtained. In general, δt is of the order of 1 fs, and from tens of thousands of such iterations the motion of the protein complex over a period of 10–1000 ps can be followed to find the minimum energy. The disadvantage of this approach is that, for an aqueous system, it requires a prohibitively long time to carry out the calculations even with modern computers. To circumvent this problem, the aqueous environment can be approximated to by using a radially dependent dielectric constant.

Molecular dynamics can be used to calculate the kinetic, potential, and total energy of the protein (P), the surfactant (S), and the complex PS. The potential energy per ligand bound (E_ν^{PE}) can, for example, be calculated from the equation

$$E_\nu^{PE} = (E_{PS_\nu}^{PE} - EP_p^{PE} - \nu E_S^{PE})/\nu \tag{6.40}$$

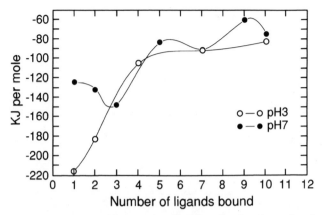

Fig. 6.17. Potential energies of interaction from molecular dynamic simulations of lysozyme with sodium n-dodecylsulphate (SDS) as a function of the number of ligands bound at pH 3 ($\bigcirc$) and pH 7 ($\bullet$) (Chen et al., 1994).

Figure 6.17 shows plots E_ν^{PE} as a function of ν for lysozyme-SDS complexes with up to 10 bound ligands. The first 2 ligands were placed on the sites found from X-ray diffraction studies (Yonath et al., 1977), after which sites were chosen on the basis of the locations of areas of high positive charge density on the protein surface. The general trend of E_ν^{PE} with increasing ν reflects a decreasing strength of binding similar to that found experimentally from the plots of ΔG vs. $\bar{\nu}$, although the magnitudes of E_ν^{PE} and $\Delta G_{\bar{\nu}}$ are very different, most likely because entropic factors are neglected in the calculations of $E_{\bar{\nu}}^{PE}$.

It is also possible to calculate the root-mean-square (RMS) displacement between the average predicted positions of the Cα atoms in protein-surfactant complexes and the corresponding atoms in the native protein without bound surfactant. Figure 6.18 shows the RMS displacements for ribonuclease A and lysozyme as a function of $\bar{\nu}$. The RMS values when $\nu = 0$ arise from the thermal motion of the structures. Ribonuclease A and lysozyme are globular proteins of similar size having molecular masses of 13,682 and 14,306, respectively; both have a single polypeptide chain with an N-terminal lysine, four disulphide bonds, and 18 cationic residues. Although they resemble each other in these structural features, the variations of RMS displacement with ν are significantly different. The slopes of the plots in Figure 6.18 are 0.43 $\pm$ 0.11 for ribonuclease A and 0.12 $\pm$ 0.03 for lysozyme. The molecular dynamic calculations thus predict that ribonuclease A is more susceptible to SDS-induced conformational changes than is lysozyme. Interestingly, this prediction is consistent with experimental observations of these proteins, specifically with the enthalpies of SDS interaction, which differ in both sign and magnitude (Fig. 6.19). At pH 3.2, the enthalpy of interaction of lysozyme with SDS is exothermic and largely arises from the binding of SDS molecules to the surface of the native molecule; only after these sites ($\sim$ 18) are saturated does the

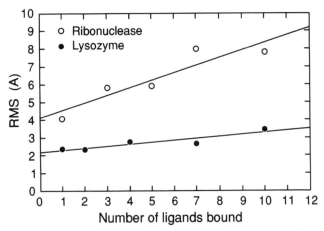

Fig. 6.18. The root-mean-square (RMS) distance between equivalent Cα atoms of the native structure of lysozyme of ribonuclease A and the appropriate structure with bound SDS after the proteins have been aligned by minimizing the total Cα distance between them using a least-squares fitting procedure: lysozyme (○), ribonuclease A (●). From Jones and Brass (1991), with permission of the Royal Society of Chemistry.

lysozyme molecule unfold (Fig. 6.19). In contrast, the enthalpy of interaction of SDS with ribonuclease A is endothermic and is dominated by the endothermicity of the conformational changes that accompany binding. Direct evidence for the existence of both exothermic and endothermic contributions comes from the microcalorimetry thermograms that show an initial exothermic signal due to binding, followed by an endotherm arising from protein unfolding (Jones et al., 1973).

Molecular dynamics can be used to give some indication of the way in which

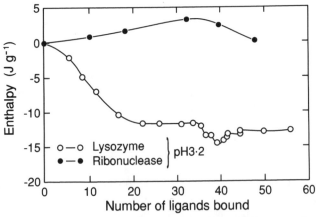

Fig. 6.19. Experimental enthalpies of interaction between lysozyme (○) and ribonuclease A (●) with SDS at pH 3.2. From Jones and Brass (1991), with permission of the Royal Society of Chemistry.

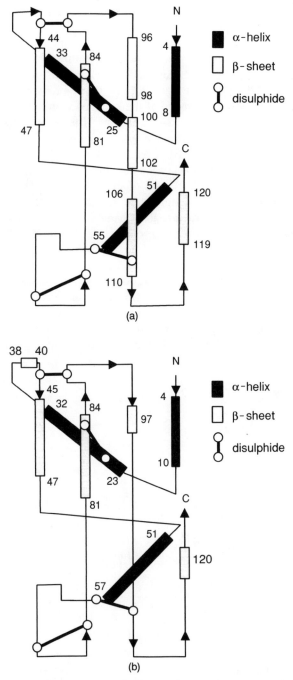

Fig. 6.20. Schematic representation of the secondary and supersecondary structure of ribonuclease A. The amino and residue numbers marked on the structure refer to the regions that are most affected by SDS binding (see text). **a.** Equilibration at pH 7. **b.** The structure at pH 7 with 10 SDS bound to the surface (Chen et al., 1994).

the interaction of a surfactant changes the structure of a protein. Figure 6.20 shows a schematic comparison between the native structure of ribonuclease A at pH 7 and the corresponding structure after the binding of 10 SDS molecules. The most significant changes induced by SDS binding are the destabilization of the β-structure—specifically the loss of β-sheet conformation between residues 96–98, 100–102, 106–110, and 119–120, and the stabilization of the α-helical structure—specifically the extension of the residues with helical conformation, e.g., 25–33 to 23–32, 51–55 to 51–57, and the N terminal helix 4–8 to 4–10. The changes in the positions of the Cα atoms relative to their positions in the native structure on binding 10 SDS molecules at pH 3 and 7 are shown in Figure 6.21. There are 10 regions of the structure in which the Cα atoms move by more than 6 Å (labeled A–J), and of these 5 regions (B, E, F, I, and J) change by more than 8 Å. It is significant that there is a considerable overlap between the regions of the structure that change most on interaction with SDS and those that are believed to change on thermal unfolding (Sheraga, 1980); these are labeled II to VI in Figure 6.21 [the first conformation change (I) involves the movement of only a single residue (tyrosine 92)]. This correspondence suggests that the pathways of SDS-induced unfolding and thermal unfolding may be similar. At the present time, the application of molecular dynamic methods to the study of surfactant-induced protein unfolding is in its infancy, but it will most likely prove to be a useful approach toward understanding the nature of surfactant and lipid interaction with proteins in the future.

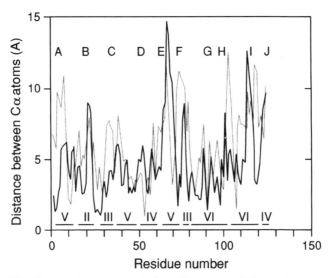

Fig. 6.21. The distance between equivalent Cα atoms of the native structure of Ribonuclease A and the structure with bound SDS after the proteins have first been aligned by minimizing the total Cα distance between them, using a least-squares fitting procedure as a function of position in the amino acid sequence. (————), results at pH 7; (.....), results at pH 3. The roman numerals mark the regions that unfold on thermal denaturation, with region II unfolding first and region VI last (Chen et al., 1994).

REFERENCES AND FURTHER READING

Alonso, A., Urbaneja, M.A., Goñi, F.M., Carmona, F.G., Canovas, F.G. and Gomez-Fernandez, J.C. (1987) Kinetic studies on the interaction of phosphatidylcholine liposomes with Triton X-100. Biochim. Biophys. Acta *902*, 237–246.

Beyhl, F.E. (1986) Interactions of detergents with microsomal enzymes: Effects of some ionic and nonionic detergents on glucose-6-phosphatase. IRCS Med. Sci. *14*, 417–418.

Chen, S.-H. and Teixeira, J. (1986) Structure and fractal dimension of protein-detergent complexes. Phys. Rev. Lett. *57*, 2583–2586.

Chen, Y., Brass, A. and Jones, M.N. (1994) Molecular modelling and thermodynamic studies of the interaction between proteins and surfactant molecules at low binding levels. Unpublished work.

Cordoba, J., Reboiras, M.D. and Jones, M.N. (1988) Interaction of n-octyl-β-D-glucopyranoside with globular proteins in aqueous solution. Int. J. Biol. Macromol. *10*, 270–276.

El-Sayert, M.Y. and Roberts, M.F. (1985) Charged detergents enhance the activity of phospholipase C (Bacillus cereus) towards micellar short-chain phosphatidylcholine. Biochim. Biophys. Acta *831*, 133–141.

Feldman, H.A. (1983) Statistical limits in Scatchard analysis. J. Biol. Chem. *258*, 12865–12867.

Finn, A., Jones, M.N., Manley, P., Paz Andrade, M.I. and Nuñez Regueira, L. (1984) Thermochemical studies on the interaction between sodium n-dodecylsulphate and catalases. Int. J. Biol. Macromol. *6*, 284–290.

Hill, A.V. (1914) The combination of haemoglobin with oxygen and with carbon monoxide. Biochem. J. *1*, 471–480.

Ibel, K., May, R.P., Kirschner, K., Szadkowski, H., Mascher, E. and Lundahl, P. (1990) Protein-decorated micelle structure of sodium dodecylsulphate: protein complexes as determined by neutron scattering. Eur. J. Biochem. *190*, 311–318.

Jones, M.N. (1988) in Biochemical Thermodynamics. M.N. Jones, Ed. Elsevier, Amsterdam, chap. 5.

Jones, M.N. (1992) Surfactant interactions with biomembranes and proteins. Chem. Soc. Rev. *21*, 127–136.

Jones, M.N. and Brass, A. (1991) in Interactions between Small Amphipathic Molecules and Proteins in Food Polymers, Gels and Colloids. E. Dickinson, Ed. Roy. Soc. Chem. special pub. no. 82, pp. 65–80.

Jones, M.N. and Manley, P. (1979) The binding of n-alkyl sulphates to lysozyme in aqueous solution. J. Chem. Soc. (Faraday Trans. 1) *75*, 1736–1744.

Jones, M.N. and Manley, P. (1980) The interaction between lysozyme and n-alkylsulphates in aqueous solution. J. Chem. Soc. (Faraday Trans. 1) *76*, 654–664.

Jones, M.N. and Manley, P. (1981) Relationship between proton and surfactant binding to lysozyme in aqueous solution. J. Chem. Soc. (Faraday Trans. 1) *77*, 827–835.

Jones, M.N. and Manley, P. (1982) The binding of sodium n-dodecylsulphate and protons to catalase. Int. J. Biol. Macromol. *4*, 201–206.

Jones, M.N. and Midgley, P.J.W. (1984) Light scattering from detergent complexed biological macromolecules. Biochem. J. *219*, 875–881.

Jones, M.N., Skinner, H.A., Tipping, E. and Wilkinson, A.E. (1972) The interaction between ribonuclease A and surfactants. Biochem. J. *135*, 231–236.

Jones, M.N., Skinner, H.A. and Tipping, E. (1975) The interaction between bovine serum albumin and surfactants. Biochem. J. *147*, 229–234.

Jones, M.N., Manley, P., Midgley, P.J.W. and Wilkinson, A.E. (1982a) The dissociation of bovine and bacterial catalase by sodium n-dodecylsulphate. Biopolymers *21*, 1435–1451.

Jones, M.N., Manley, P. and Wilkinson, A.E. (1982b) The dissociation of glucose oxidase by sodium n-dodecylsulphate. Biochem. J. *203*, 285–291.

Jones, M.N., Manley, P. and Holt, A. (1984) Cooperativity and effect of ionic strength on the binding of sodium n-dodecylsulphate to lysozyme. Int. J. Biol. Macromol. *6*, 65–68.

Jones, M.N., Finn, A., Mosavi-Movahedi, A. and Waller, B.J. (1987) The activation of *Aspergillus niger* catalase by sodium n-dodecylsulphate. Biochim. Biophys. Acta *913*, 395–398.

Kale, K., Kresheck, G.C. and Vandekooi, G. (1978) A calorimetric comparison of the interaction of sodium dodecylsulfate with cytochrome c and erythrocyte glycoproteins. Biochim. Biophys. Acta *535*, 334–341.

Kirkpatrick, F.H., Gordesky, S.E. and Marinetti, G.V. (1974) Differential solubilization of proteins, phospholipids and cholesterol of erythrocyte membranes by detergents. Biochim. Biophys. Acta *345*, 154–161.

Klotz, I.M. (1982) Numbers of receptor sites from Scatchard graphs: Facts and fantasies. Science *217*, 1247–1248.

Klotz, I.M. and Hunston, D.L. (1984) Mathematical models for ligand-receptor binding: Real sites, ghost sites. J. Biol. Chem. *259*, 10060–10062.

Lichtenberg, D. (1985) Characterisation of the solubilization of lipid bilayers by surfactants. Biochim. Biophys. Acta *821*, 470–478.

Lichtenberg, D., Robson, R.J. and Dennis, E.A. (1983) Solubilization of phospholipids by detergents: Structural and kinetic aspects. Biochim. Biophys. Acta *737*, 285–304.

Lundahl, P., Mascher, E., Kameyama, K. and Takagi, T. (1990) Water-soluble proteins do not bind octyl glucoside as judged by molecular sieve chromatographic techniques. J. Chromatogr. *518*, 111–121.

Makino, S., Maezawa, R., Moriyama, R. and Takagi, T. (1986) Determination of polypeptide chain molecular weights of human and bovine band 3 protein from erythrocyte membranes by low-angle laser light scattering combined with high-performance gel chromatography in the presence of sodium dodecyl sulphate. Biochim. Biophys. Acta *874*, 216–219.

McCammon, J.A. and Harvey, S.C. (1987) Dynamics of Proteins and Nucleic Acids. Cambridge University Press, Cambridge.

Nelson, C.A. (1971) The binding of detergents to proteins. Part 1: The maximum amount of dodecyl sulphate bound to proteins and the resistance to binding of several proteins. J. Biol. Chem. *246*, 3895–3901.

Partearroyo, M.A., Pilling, S.J. and Jones, M.N. (1991) The lysis of isolated fish (*Oncorhynclus mykiss*) gill epithelial cells by surfactants. Comp. Biochem. Physiol. *100C*, 381–388.

Pfeil, W. and Privalov, P.L. (1976) Thermodynamic investigations of proteins. Part 3: Thermodynamic description of lysozyme. Biophys. Chem. *4*, 41–50.

Reboiras, M.D. and Jones, M.N. (1982) Detection of specific serum albumin–sodium n-dodecylsulphate complexes by polyacrylamide gel eletrophoresis. Electrophoresis 3, 371–321.

Ruiz, J., Goñi, F.M. and Alonso, A. (1988) Surfactant-induced release of liposomal contents: A survey of methods and results. Biochim. Biophys. Acta 937, 127–134.

Scatchard, G. (1949) The attraction of proteins for small molecules and ions. Ann. NY Acad. of Sci. 51, 660–672.

Schwarz, G. (1976) Some general aspects regarding the interpretation of binding data by means of a Scatchard plot. Biophys. Struct. Med. 2, 1–12.

Sheraga, A.S. (1980) in Protein Folding (R. Jaenicke, Ed.) Elsevier North–Holland Biomedical Press, Amsterdam, pp. 261–288.

Shiao, Y.-J., Chen, J.-C. and Wang, C.-T. (1989) The solubilization and morphogical change of human platelets in various detergents. Biochim. Biophys. Acta 980, 56–58.

Snel, M.M.E., Kaptein, R. and de Kruijff, B. (1991) Interaction of apocytochrome c and derived polypeptide fragments with sodium dodecylsulphate micelles monitored by photochemically induced dynamic nuclear polarization ^{1}H NMR and fluorescence spectroscopy. Biochemistry 30, 3387–3395.

Tipping, E., Jones, M.N. and Skinner, H.A. (1974) Enthalpy of interaction between some globular proteins and sodium n-dodecylsulphate in aqueous solution. J. Chem. Soc. (Faraday Trans. 1) 70, 1306–1315.

Urbaneja, M.A., Nieva, J.L., Goñi, F.M. and Alonso, A. (1987) The influence of membrane composition on the solubilizing effects of Triton X-100. Biochim. Biophys. Acta 904, 337–345.

Urbaneja, M.A., Goñi, F.M. and Alonso, A. (1988) Structural changes induced by Triton X-100 in sonicated phosphatidylcholine liposomes. Eur. J. Biochem. 173, 585–588.

Weber, K. and Osborn, M. (1975) in The Proteins. (H. Neurath, R.L. Hill, C.-L. Boeden, Eds.) Vol. 1. 3rd ed. Academic Press, San Diego, pp. 179–223.

Wyman, J. (1965) The binding potential: A neglected linkage concept. J. Mol. Biol. 11, 631–644.

Yonath, A., Podjamy, A., Honig, B., Sielecki, A. and Traub, W. (1977) Crystallographic studies of protein denaturation and renaturation Part 2: Sodium dodecyl sulphate induced structural changes in triclinic lysozyme. Biochemistry 16, 1418–1424.

CHAPTER 7

MEMBRANE STRUCTURE

7.1. THE COMPOSITION OF BIOMEMBRANES

7.1.1. Lipids

The lipid matrix of biomembranes contains a variety of lipids. These are often based upon glycerol or sphingosine (sphingolipids). The structures of some of the common phospholipids were shown in Chapter 1; others are shown in Figure 7.1.

Some unusual lipid molecules occur in thermophilic prokaryotic organisms which may be either eubacteria or archaebacteria. Thermophilic eubacteria belong to different genera that comprise a wide variety of microorganisms— aerobic, anaerobic, spore-forming, Gram-positive, Gram-negative—capable of growing over a wide range of pH (2.0–9.0).

The polar lipids from thermophilic eubacteria are based either on ester (genera *Thermus* and *Bacillus*) or on ether (genus *Thermodesulfotobacterium*) derivatives of glycerol with the 1,2-*sn* configuration. The predominant acyl components are saturated fatty acids with a variable number of carbons atoms; moreover, most of the fatty acids are branched. In some species of the *Bacillus* genus, ω-cyclohexyl fatty acids are also present. In ether lipids, the alkyl components are iso- and anteiso-saturated long-chain alcohols. Often the relative abundance of the different fatty acids varies with the composition of the medium, temperature of growth, age of the culture, and strain. Usually the proportion of the longer chains and/or the ratio iso/anteiso increases with temperature, in line with the melting temperature of the chains. Cells of some species respond to a rise in the temperature by increasing their lipid content (in either the glycolipid or the phospholipid fraction, or in both).

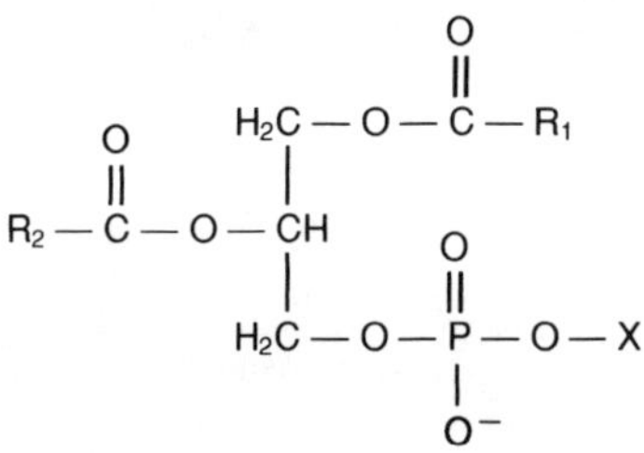

$$
\begin{array}{c}
\quad\quad\quad\quad\quad\quad\quad O \\
\quad\quad\quad\quad\quad\quad\quad \| \\
\quad\quad\quad H_2C-O-C-R_1 \\
O\quad\quad\quad | \\
\| \quad\quad\quad | \\
R_2-C-O-CH \quad\quad\quad O \\
\quad\quad\quad | \quad\quad\quad\quad \| \\
\quad\quad\quad H_2C-O-P-O-X \\
\quad\quad\quad\quad\quad\quad | \\
\quad\quad\quad\quad\quad\quad O^-
\end{array}
$$

where

$X = -H$ phosphatidic acid   (I)

$= -CH_2-CH_2-N^+(CH_3)_3$ phosphatidyl choline or lecithin   (II)

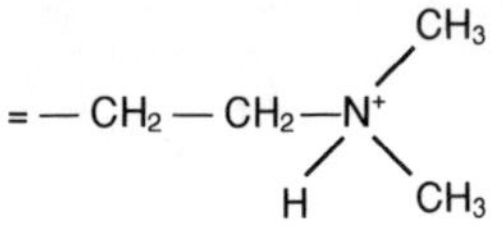

$= -CH_2-CH_2-N^+\begin{smallmatrix}CH_3\\ \\ CH_3\end{smallmatrix}$ (with H)   phosphatidyl 1 (N - dimethyl) ethanolamine   (III)

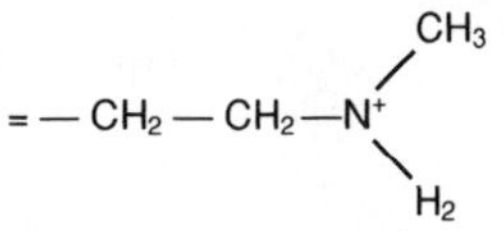

$= -CH_2-CH_2-N^+\begin{smallmatrix}CH_3\\ \\ H_2\end{smallmatrix}$ phosphatidyl 1 (N - methyl) ethanolamine   (IV)

$= -CH_2-CH_2-N^+H_3$ phosphatidyl ethanolamine   (V)

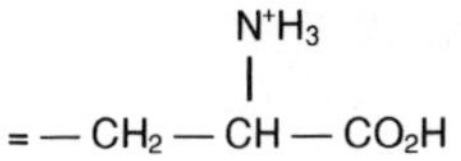

$= -CH_2-\underset{\underset{}{\overset{\overset{N^+H_3}{|}}{CH}}}{}-CO_2H$ phosphatidyl serine   (VI)

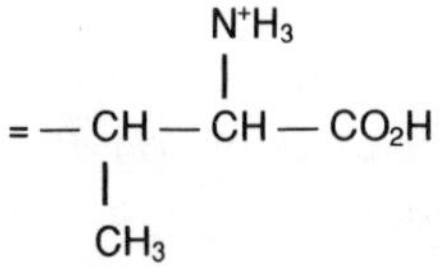

$= -\underset{\underset{CH_3}{|}}{CH}-\underset{\overset{N^+H_3}{|}}{CH}-CO_2H$ phosphatidyl threonine   (VII)

(a)

Fig. 7.1. Structures of some common phospholipids.

Archaebacteria constitute a novel family of prokaryotic microorganisms, recently identified as the third kingdom of life. Archaebacteria are found in exceptional ecological niches and are classified—according to their habitat— into halophiles, thermophiles, and methanogens. The distinction between eu- bacteria and archaebacteria is based upon phylogenetic criteria—sequence of ribosomal RNAs, subunit composition of RNA polymerase, size and shape of ribosomes—or upon the chemical composition of the lipids.

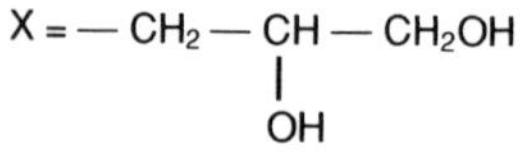

$$X = - CH_2 - CH - CH_2OH \qquad \text{phosphatidyl glycerol} \qquad \text{(VIII)}$$ 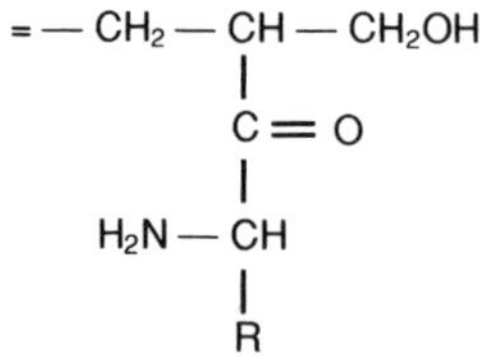
$$\qquad\qquad | $$
$$\qquad\qquad OH$$

$$= - CH_2 - CH - CH_2OH \qquad \text{o - amino acid ester of} \qquad \text{(IX)}$$
$$\qquad\quad | \qquad\qquad\qquad \text{phosphatidyl glycerol}$$
$$\qquad\quad C = O$$
$$\qquad\quad |$$
$$\quad H_2N - CH$$
$$\qquad\qquad |$$
$$\qquad\qquad R$$

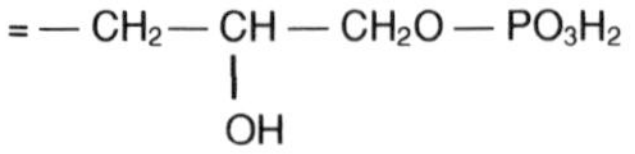

$$= - CH_2 - CH - CH_2O - PO_3H_2 \qquad \text{phosphatidyl} \qquad \text{(X)}$$
$$\qquad\quad | \qquad\qquad\qquad\qquad \text{1 glycerophosphate}$$
$$\qquad\quad OH$$

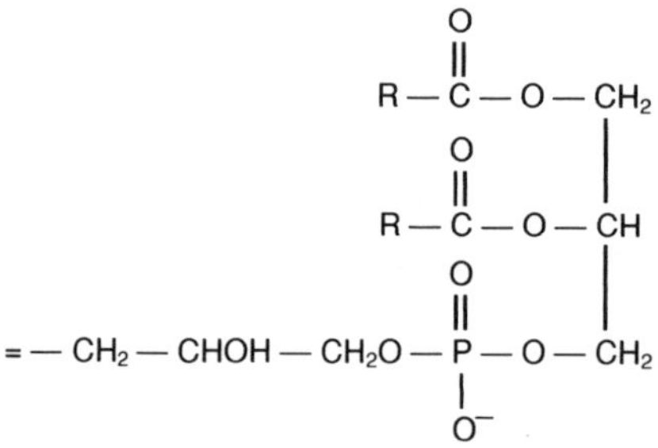

$$
\begin{array}{l}
\qquad\qquad\qquad O \\
\qquad\qquad\qquad || \\
\qquad\qquad R - C - O - CH_2 \\
\qquad\qquad\qquad O \qquad\qquad | \qquad \text{diphosphatidyl} - \text{(XI)} \\
\qquad\qquad\qquad || \qquad\quad | \qquad \text{glycerol} \\
\qquad\qquad R - C - O - CH \\
\qquad\qquad\qquad O \qquad\qquad | \\
\qquad\qquad\qquad || \qquad\quad | \\
= - CH_2 - CHOH - CH_2O - P - O - CH_2 \\
\qquad\qquad\qquad\qquad\qquad | \\
\qquad\qquad\qquad\qquad\qquad O^-
\end{array}
$$

(b)

Fig. 7.1. (*Continued*)

The lipids extracted from archaebacteria are characterized by remarkable structural features. First, these lipids are based exclusively on *ether* linkages; second, the molecules are formed by condensation of glycerol or of more complex polyols with isoprenoid alcohols containing 20, 25, or 40 carbon atoms; third, the glycerol ethers contain a 2,3-di-O-*sn*-glycerol (Luzzati et al., 1986).

In high-temperature organisms, the chemical structure of the polar lipids is more strongly correlated with the primary kingdom to which the microorganism belongs (eubacteria vs. archaebacteria) than with the temperature of growth. The lipids from thermophilic eubacteria share a variety of chemical features with the lipids from mesophilic eubacteria (and from eukaryotes)—glycerol in the *sn*-1,2 configuration, hydrocarbon chains of the fatty acid-type—whereas the lipids from archaebacteria (both thermophilic and mesophilic) are all based upon isoprenoid hydrocarbon chains, ether linked to glycerol, whose configuration is 2,3-di-O-*sn*.

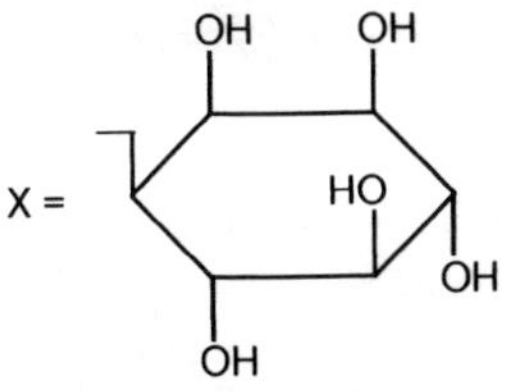

phosphatidyl 1 (myo) –
inositol
or monophosphoinositide   (XII)

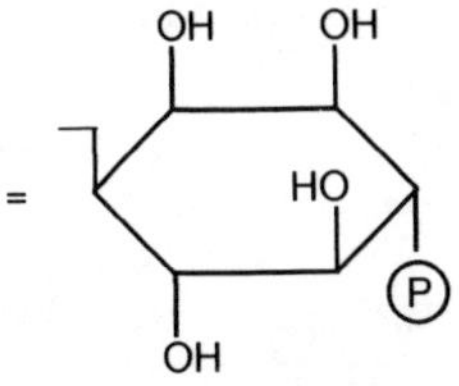

phosphatidyl 1 (myo) –
inositol – 4 – phosphate
or diphosphoinositide   (XIII)

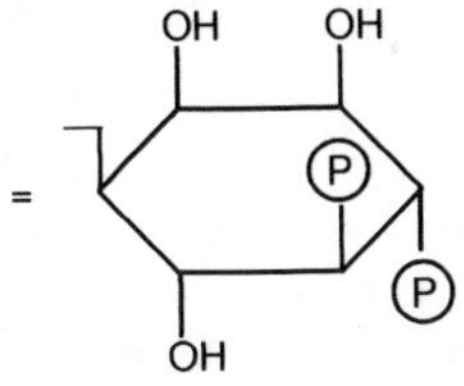

phosphatidyl 1 (myo) –
inositol – 4 – phosphate
or triphosphoinositide   (XIV)

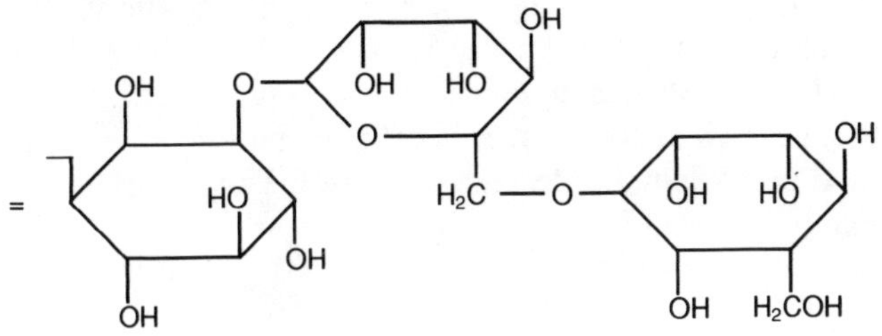

phosphatidyl 1 (myo) inositoldimannoside   (XV)

(c)

Fig. 7.1. (*Continued*)

7.1.2. Proteins

The ratio of protein to lipid present in a biomembrane ranges considerably from one cell system to another. It is high in the purple membrane with the protein bacteriorhodopsin and lower in the myelin membranes (see Table 7.1). Investigations into membrane proteins led to a broad division of these molecules into two classes: integral (intrinsic) and peripheral (extrinsic) membrane proteins. The latter are easily dissociated from the membrane, often with little or no effect on either the structure or function of the protein. Thus they behave much like cytoplasmic proteins. Integral membrane proteins are intimately as-

TABLE 7.1. Composition of Various Membranes Expressed as Their Protein-Lipid Weight

Membrane	W
Bovine myelin	0.28
Human erythrocyte	1.11
Outer mitochondrial membrane	1.22
Retinal rod outer segment	1.44
Hepatocyte plasma membrane	1.50
Micrococcus lysodeikticus	1.91
Saccharomyces cerevisiae plasma membrane	1.98
Chloroplast lamella	2.22
Acholeplasma	2.33
Halophilic bacteria	3.14

sociated with the hydrophobic interior of the lipid bilayer; their dissociation from membranes requires complete physical disruption of the membrane.

The integral proteins contain hydrophobic regions, whereas the extrinsic proteins resemble soluble enzymes. The integral proteins are synthesized somewhat differently, with signal or address sequences, depending on their final location. The integral proteins are concerned with catalytic functions such as membrane transport.

Animal and some plant and fungal cells contain sugar residues on their surface. These have an important role in cell recognition processes, in antigenic properties, and in hormone reception. Some of the sugar residues are attached to lipids, mainly sphingolipids. A great number are covalently linked to proteins. In some animal cells, this is the substantial part of the so-called glycocalyx.

The sugars are either O-linked or N-linked. The first type bind via serine or threonine (and, exceptionally, hydroxylysine); the second type do so via asparagine. The O-linked glycoproteins obligatorily contain N-acetyl-D-galactosamine (GalNAc) as the first and often the second sugar; the N-linked ones always bind to N-acetyl-D-glucosamine (GlcNAc) and always contain the sequence

$$\text{Man} \xrightarrow{1,4} \text{GlcNAc} \xrightarrow{1,4} \text{GlcNAc} \longrightarrow \text{Asn}$$

(Man D-mannose; Fuc D-fucose; Gal D-galactose; SA sialic or N-acetyl-neuraminic acid)

A typical glycoprotein is glycophorin A of erythrocytes with 131 amino acid residues plus a full 60% of total weight in sugar residues, 15 of the oligosaccharide chains O-linked, one N-linked.

7.2. THE ORIGIN OF THE BILAYER CONCEPT IN MEMBRANES

The first suggestion that a lipid bilayer was the core of a biological membrane was proposed by Gorter and Grendel (1925). These workers extracted the lipid from erythrocyte membranes, compressed it at an air–water interface, and showed that it occupied a surface area equal to twice the external area of the cells. (Later workers showed that there were some compensating errors in these deductions (Bar et al., 1966).) Danielli and Davson (1935) developed this model to include proteins, but suggested that the lipid polar groups were coated with layers of protein in order to account for the low membrane tension. The development of the electron microscope appeared to support this concept, giving rise to the "unit membrane" hypothesis. At this time electron microscope studies required a process of fixation, often with osmium tetroxide, dehydration, and embedding, etc., and a particular difficulty was to be sure of the location of the osmium material. Other workers argued against this model, emphasizing the high amount of protein present in certain membrane systems. Work on myelin using X-ray methods also appeared to fit the unit membrane structure.

The present accepted view is that the phospholipids in biomembranes are organized in a bilayer structure (the archaebacteria are considered to be an exception to this rule and have only a single lipid spanning the lipid structure). Perhaps the best proof of the existence of a lipid bilayer structure is based upon the application of freeze fracture electron microscopy. The fracture takes place along the center of the bilayer, i.e., the fracture plane, and particles can be seen representing the embedded integral proteins. There have been many suggestions and attempts to demonstrate nonlamellar structures as permanent structures in biomembranes, but this is not yet proven. (Lipids extracted from biomembranes often form hexagonal or cubic phases at physiological temperatures, but this does not mean that these are present in natural membranes when protein is present.) Within the bilayer, the rate of lateral diffusion can be fast (10^7 sec^{-1}) while the rate of flip-flop can be slow, hours to days (see Chap. 10).

7.3. THE FLUID-MOSAIC MODEL

The fluid-mosaic model proposed by Singer and Nicolson (1972) was based upon the work of a number of scientists. The model emphasized the concept of a fluid lipid sheet of varying composition and fluidity. The concept of lipid membrane fluidity had been previously proposed by Chapman and coworkers (1966). Embedded in the sheet are the integral proteins, which are able to undergo lateral and rotational diffusion. The integral proteins were suggested to have α-helical structures.

This model was a useful summary of scientific studies of many workers in the biomembrane field. There are of course exceptions to the model; e.g., the bacteriorhodopsin protein is fixed in position within the lipid sheet and does not undergo rotational and lateral diffusion. The lipids are also particularly rigid and immobile. Some biomembranes contain regions of crystalline lipid and fluid lipid, e.g., *Acholeplasma laidlawii* membranes. Some membrane proteins have been shown to have β-barrel structure, e.g., porin (Cowan and Rosenbusch, 1994).

7.4. STATIC AND DYNAMIC LIPID ASYMMETRY IN CELL MEMBRANES

The asymmetric distribution of lipid classes was first observed in human erythrocyte membranes by Bretscher (1972). Most biological membranes appear to have a different phospholipid composition in their inner and outer leaflets. In erythrocytes, phosphatidylserine (PS) phosphatidylethanolamine (PE), and probably phosphatidylinositol (PI) are located mainly in the inner layer, while phosphatidylcholine (PC) and sphingomyelin (SM) are essentially in the outer layer.

A number of techniques have been used to determine lipid asymmetry in biomembranes. These techniques include chemical labeling with nonpenetrating agents—e.g., with trinitrobenzenesulphonic acid (TNBS) or fluoresceinamine—immunological methods, phospholipase digestion of membrane phospholipids, use of phospholipid-exchange proteins, and physicochemical methods such as X-ray diffraction and NMR spectroscopy.

Variations of the order of 10% exist between the different reports dealing with lipid asymmetry in normal erythrocytes. Abnormal distributions have been reported for aged red cells, sickle cells, malaria-infected red cells, etc. It has been reported that the average fatty acid composition of PS and PE shows more unsaturation than for PC and SM. Within the same class of phospholipids (SM or PE), acyl chains from the outer layer differ from those of the inner layer.

There have been many attempts to investigate lipid asymmetry in internal membranes from eukaryotic cells. Unfortunately, when the same systems have been studied by different laboratories, the results often differ considerably. This may be due to the very rapid lipid redistribution within each membrane and from membrane to membrane within each eukaryotic cell.

Very little is known concerning lipid asymmetry in plant cells. Reports have appeared concerning the topology of lipids in bacteria and viruses. The lipid asymmetry in viruses most likely reflects the asymmetry of the infected cells. The appearance of PS on the outer monolayer of a cell membrane stimulates protein activity. The presence of PS produces more than a millionfold increase in the rate of thrombin formation. The conversion of prothrombin into thrombin is followed by the formation of fibrin strands and then the clot. The main function of the platelet membrane in this process is to provide a catalytic surface on which the coagulation factors interact, thereby increasing their local concentration. In fact, intravascular clotting can be triggered by PS exposure on the outer surface of red cells as well; this may happen *in vivo* in the crisis phase of sickle cell disease.

It has been shown that lipids which exhibit weak acid or weak base characteristics, such as free fatty acids, rapidly redistribute across liposome lipid bilayers in response to a transmembrane pH gradient. Certain phospholipids, such as PG and PA, can also cross a lipid bilayer under the influence of a pH gradient. The transmembrane movement of PG occurs via permeation of the uncharged (protonated) dehydrated form, which can exhibit half-times of trans-bilayer transport of the order of seconds.

Lipid asymmetry was thought to be the consequence of asymmetrical membrane biogenesis and asymmetrical lipid turnover by endogenous phospholipases and reacylases, together with the asymmetrical insertion of lipid constituents. The differences in potential and/or pH between the two surfaces could also explain the stability of the asymmetrical distribution. However, in 1984, the existence of an ATP-requiring mechanism responsible for the specific translocation of aminophospholipids (PS and PE) was demonstrated in human red cells and later in other plasma membranes (see Devaux, 1991).

7.5. PROTEIN AND GLYCOPROTEIN ARRANGEMENTS

The proteins associated with biomembranes can be arranged in different ways with respect to the lipid bilayer matrix. Some of these arrangements are shown in Figure 7.2.

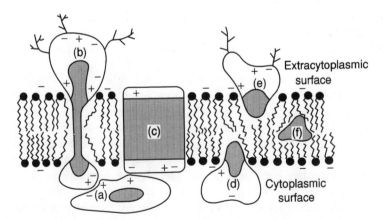

Fig. 7.2. Topography of membrane proteins showing peripheral (extrinsic) proteins (a) and integral (intrinsic) proteins (b–f). From Houslay and Stanley (1982). Copyright 1982, John Wiley & Sons.

7.5.1. Integral Proteins

In a few cases, the protein is fully submerged in the lipid matrix—e.g., in myelin—while integral membrane proteins that span the lipid matrix are abundant, particularly those associated with transport and energy transduction. In some cases, the membrane-spanning helical polypeptide arrangements have been directly demonstrated using electron diffraction techniques as with bacteriorhodopsin. In other cases, hydropathy plots have been used to create models, giving rise to predictions of a number of spanning segments (see Chapter 1).

Transduction proteins that activate G proteins, such as rhodopsin, are considered to have seven spanning helices, but a range of these spanning segments has been postulated. The polypeptide arrangements within the lipid matrix are often in the form of α-helices. This has been shown with a variety of proteins using FTIR spectroscopy (Harris and Chapman, 1992), but in the case of porin a β-barrel arrangement occurs (Cowan and Rosenbusch, 1994).

7.5.2. Extrinsic Proteins

The extrinsic proteins can be removed from the biomembrane structure by manipulating the ionic strength; e.g., with the erythrocyte membrane, exposure to low salt and a chelating agent removes the extrinsic protein spectrin. Other extrinsic proteins include those located on the outer surface of plasma membranes and are anchored to the surface by a phospholipase-sensitive lipid component. These anchor-type proteins are discussed in more detail in Chapter 9.

Another method used to provide information on the topology of membrane proteins is the method of using site-directed antibodies. This method arises from the fact that antibodies raised against short synthetic peptides frequently recognize the same sequence in native proteins. Thus antibodies can be produced against proteins of known sequence even when they have not been isolated. These site-directed antibodies are commonly raised in the form of polyclonol sera.

In proteins of known three-dimensional structure, the best sequences to give useful antibodies (i.e., which bind to the native proteins) appear to be those with the greatest segmental mobility. Typically peptides of 15–20 residues in length are chosen for the production of site-directed antibodies. It is usually necessary to couple the peptide to a carrier protein—such as keyhole limpet haemocyanin—to obtain a good immune response (say, with rabbits).

This technique has been applied for probing the topology of the human erythrocyte glucose transport protein (Davies et al., 1990). Antibodies raised against peptides from a large central extramembranous loop recognize the native protein. By examining the ability of antibodies to bind to right-side-out and inside-out membrane vesicles, these workers were able to show which sequence is located on the cytoplasmic side of the membrane, leading to the model shown in Figure 7.3.

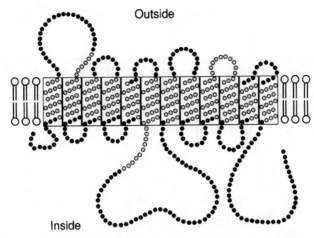

Outside

Inside

Fig. 7.3. Location within the glucose transporter sequence of synthetic peptides (shown in black) against which antibodies have been raised. After Davies et al. (1990).

7.6. ION CHANNEL STRUCTURES AND RECEPTOR PROTEINS

There is considerable interest in the structure of voltage-gated ion channels (MacKinnon and Miller, 1989). These ion channels are the plasma proteins responsible for the propagation of electrical signals in excitable cells such as nerve and muscle. There are three types of voltage-gated channels, namely, K^+, Na^+, and Ca^{2+}. A typical K^+ channel is considered to be built as a tetramer of a 600-residue subunit. The Na^+ and Ca^{2+} channels are formed from a single polypeptide of about 2,000 residues containing four homologous domains.

The use of molecular biology techniques—including site-directed mutagenesis and probe molecules that bind specifically to the pores of K^+ channels—has led to a model shown in Figure 7.4. Three types of ligands were known to interact specifically with the pore for K^+ ions: scorpion venom peptides, tetraethylammonium (TEA), and the ions themselves. A localized region linking the fifth and sixth membrane-spanning helical stretches of amino acids (S_5 and S_6) is thought to form the K^+ channel pore.

7.6.1. Receptor Proteins

Many transmembrane signaling systems consist of three membrane-bound protein components: (a) a cell surface receptor; (b) an effector such as an ion channel or the enzyme adenylate cyclase; and (c) a guanine nucleotide-binding regulatory protein or G protein, which is coupled to both receptors to mediate the action of light, peptide hormones, and neurotransmitters. Nearly all G protein-coupled receptors have sequence similarity. Based upon this sequence similarity, it is thought that they have a similar topological motif—i.e., they

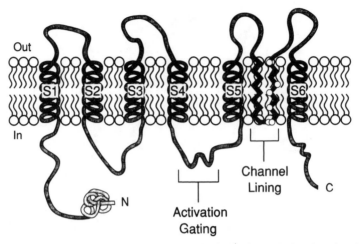

Fig. 7.4. A model of a single subunit of a typical K^+ channel showing the six membrane spanning α-helical structures $S_1 \rightarrow S_6$. The channel lining region (the pore) and the activation gating regions are shown (after Miller, 1992).

consist of seven hydrophobic (possibly α-helical segments) that span the lipid bilayer. This seven-helix motif is based upon the analogy with the structure of bacteriorhodopsin and the presence of hydrophobic sequences. A proposed model (Dohlman et al., 1987) of the folding of the β-adrenergic receptor is shown in Figure 7.5.

7.7. LIPID PHASE TRANSITIONS AND BIOMEMBRANE SYSTEMS

We have seen in Chapter 4 the phase transition behavior that occurs within phospholipid-water systems. What is the relevance of such studies to natural bio-membranes?

We have seen that biomembrane structures are built upon a lipid bilayer structure into which the proteins are inserted (Chapman, 1984). The properties of the lamellar phase of the lipid-water systems are therefore of direct relevance. The properties of this lamellar phase include movement of the lipids within the fluid bilayer and the possibility of triggering phase changes of this phase by temperature or dehydration where the lipid chains may crystallize or, in some cases, cause the production of hexagonal or cubic phases. Biological membranes of course contain complex mixtures of lipids. An additional complexity exists in natural biomembranes, in that (as we have seen) the lipid composition is normally asymmetric across the bilayer structure.

From an examination of the various lipid-water phase diagrams, it can be seen that the degree of biomembrane "fluidity" is a direct reflection of the transition temperature of the lipids; i.e., the fluidity is greatest when highly unsaturated lipids are present and less when more saturated lipids are present.

In some cases, the transition temperature of the lipids will be below or at growth temperature; in other cases, it may be above the growth temperature or encompass it. The appropriate phase transition characteristic of the lipids present therefore determines the appropriate fluidity of the particular cell membrane and also the correct lipid phase separation characteristics. These in turn can affect membrane elasticity—the insertion, aggregation, and diffusional movements of the protein and lipid components—as well as permeability characteristics.

In certain practical situations, it is important to freeze cells and tissues, as in the various requirements in the field of cryobiology, so that major changes of temperature are involved. When cells or membranes are frozen to very low temperatures, lipid phase characteristics and phase separation need to be considered. Whether the lipid crystallizes before the ice melting point or at a lower temperature could be important in determining freeze damage. The various phenomena of lipid phase separation and protein aggregation are relevant to this situation.

Studies have been made of permeability characteristics for water and various other molecules above and below the lipid phase transition temperature; e.g., the kinetics of water permeability through lipid systems has been studied and related to lipid fluidity. Thus there is a marked increase in water permeability as the lipid chains become more unsaturated. This is also the case with non-electrolytes such as glycerol and erythritol. The self-diffusion rate of $^{22}Na^+$ through lecithin bilayers also shows a marked increase at the transition temperature to the liquid-crystalline form. The effect of cholesterol on the lipid chains when the lipid is in the fluid condition is to decrease water permeability. Black lipid membranes also show the same effect. These results are consistent with a reduction in the fluidity characteristics of the lipid. On the other hand, the presence of cholesterol is to enhance the rate of water permeability of liposomes derived from saturated liposomes.

Triggering mechanisms may occur as a result of interactions with the biomembrane by metal ions, proteins, or drugs so that local changes of fluidity, phase separation, and hence changes of permeability characteristics can take place.

7.7.1. The Hexagonal Phase

Unsaturated fatty acids are known to be fusogenic agents, and so it was of great interest when it was found that their incorporation into phospholipid bilayers, or in natural (erythrocyte) membranes, tends to induce the formation of an H_{II} phase, as well as promote fusion. Retinol and monoolein are found to have similar effects on erythrocyte membranes. The effect of alcohols on hydrated egg PE membranes is to stabilize the L_{α} phase for short chains (ethanol, butanol), but for longer chains to tend to promote H_{II} phase formation.

Many natural membranes are rich in lipids that have strong tendencies to form H_{II} phases, and considerable effort has been spent on trying to detect such structures in biological systems. Mitochondrial inner membranes are particularly rich in cardiolipin, which forms the H_{II} phase in the presence of low levels of divalent ions.

A particular example of the relevance of these mesomorphic phases is shown by studies of *Acholeplasma laidlawii*. This is a simple, procaryotic microorganism without cell walls that possesses a number of features which make it an attractive system for studying the roles of lipids in biological membranes. An especially useful feature of this organism as a model membrane system is the ability to dramatically alter its membrane lipid fatty acid composition and cholesterol content by appropriate manipulation of the lipid composition of the growth medium. Utilizing this ability, Wieslander et al. (1980) have shown that alterations in membrane lipid fatty acid composition and cholesterol content, as well as in growth temperature, induce marked changes in the quantitative distribution of the major polar lipids on the membrane of *A. laidlawii* strain A.

These workers have postulated that in all biological membranes a certain balance of bilayer-preferring and nonbilayer-preferring lipids must be maintained in order to ensure an optimal degree of lipid stability and functionality. For *A. laidlawii*, this requires maintaining an optimal balance between the amount of monoglycosyldiacylglycerol (MGDG) and the other membrane lipids present in the face of alterations in growth temperature or in the fatty acid composition or cholesterol content of the growth medium.

There are numerous reports on model systems of membrane lipids, which exhibit periodic structures with electron microscopy textures like the cubic phases, for example, "intramembrane particles" or "lipidic particles." Extensive studies have shown that the lipid systems reported to give lipidic particles also form cubic phases.

Luzzati et al. (1986) have reported most interesting phase properties of lipids from the thermoresistant organism *Solfolobus solfataricus*. These ether lipids form cubic phases under physiological conditions, and a membrane model with protein "plugs" has been proposed.

7.8. THE STRUCTURE OF THE HUMAN ERYTHROCYTE MEMBRANE

The erythrocyte is a remarkably durable cell, surviving thousands of passes through the circulation during its 120-day life. The basis for the resilience of the cell is the mechanical properties of the plasma membrane. Direct measurements of this deformation of erythrocytes have shown that the membrane behaves like a semisolid with elastic properties. The major proteins are shown in Table 7.2.

The structural component of the erythrocyte responsible for the elastic properties is a membrane-associated assembly of proteins commonly referred to as the cytoskeleton. The cytoskeleton has been visualized as a two-dimensional

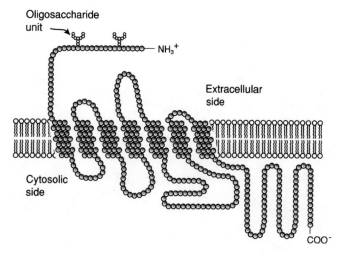

Fig. 7.5. Proposed folding of the β-adrenergic receptor. This seven-helix motif is a common feature of transmembrane receptors that activate G proteins. Transmembrane helices are shown. Two N-linked oligosaccharide units are located on the extracellular side of the plasma membrane. A loop on the cytosolic side participates in activating G proteins. Phosphorylation of multiple serine residues in the carboxyl-terminal tail prevents the receptor from interacting with the G protein. From Dohlman et al. (1987), with permission of the American Chemical Society.

meshwork of filaments about 100–140 nm in length (see Fig. 7.6). The principal protein components remaining in the cytoskeleton after high-salt extraction are spectrin, band 4.1, erythrocyte actin, and band 4.9. Table 7.2 shows the major proteins and glycoproteins of the human red cell membrane.

Advances in understanding the red cell membrane have provided the first detailed ''map'' of the protein linkages involved in attachment of actin to a plasma membrane. Actin filaments are associated in a ternary complex with band 4.1 and spectrin. Spectrin is a flexible, rod-shaped molecule 200 nm in length and forms a meshwork underlying the cytoplasmic surface of the plasma membrane. Spectrin is attached to the membrane bilayer by association with ankyrin. Ankyrin, in turn, is associated with the cytoplasmic domain of the anion transporter, which is an integral membrane protein. Each of the linking proteins has been purified, and their associations with neighboring proteins have been described quantitatively.

In recent developments, spectrin and the anion channel protein have been partially sequenced, and the domain structure of ankyrin and band 4.1 has been resolved. An additional association between spectrin and the membrane by linkage of band 4.1 to glycoproteins has been suggested (see Fig. 7.6).

TABLE 7.2. Major Proteins of the Human Red Cell Membrane*

Protein	Subunit M_1
Peripheral proteins	
Spectrin	$\alpha = 260,000$
	$\beta = 225,000$
Ankyrin	215,000
Band 4.1[a]	78,000
Band 4.2[a]	72,000
Band 4.9	45,000
Actin	43,000
Glyceraldehyde 3-phosphodehydrogenase	35,000
Band 7[a]	29,000
Band 8[a]	23,000
Tropomyosin	29,000
Integral proteins	
Band 3[a]	89,000
Glycophorin A	31,000
Glycophorin B[b]	23,000
Glycophorin C[b]	29,000
(Glycoconnectin)	

*Quinn and Chapman (1991).
[a]Nomenclature based on electrophoretic mobility.
[b]Molecular mass estimated assuming 60% carbohydrate for all three glycophorins.

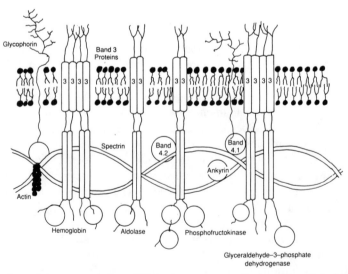

Fig. 7.6. A model representation of links between membrane and cytoskeletal components of the erthrocyte membrane. After Quinn and Chpaman (1991).

7.9. SPECTROSCOPIC STUDIES OF BIOMEMBRANES

Although cDNA sequencing has led to the elucidation of the amino acid sequence of many membrane proteins, secondary and tertiary structural information on them is very limited. X-ray diffraction methods, which have been very successful in solving the three-dimensional structure of soluble proteins, cannot be readily applied to membrane proteins. This is because it is very difficult to obtain appropriate crystals of membrane proteins for X-ray crystallographic studies. The only transmembrane proteins to produce good three-dimensional crystals suitable for crystallography have been the photosynthetic reaction centers. Other physical techniques useful for studying soluble protein structures—such as nuclear magnetic resonance (NMR) and circular dichroism (CD) spectroscopy—are also not readily applicable to the study of membrane proteins. The application of NMR spectroscopy to the study of membrane proteins has met a number of technical difficulties. The proteins are in an anisotropic environment within the lipid bilayer and often have high molecular weight.

Some of these difficulties can be overcome by a combination of methods. Thus high resolution techniques can be applied to the membrane proteins in solution in micelles and then solid-state spectroscopy to the proteins within the lipid bilayers. An example of this is recent studies made of the membrane-bound bacteriophage Pf1 coat protein. In this case extensive use was made of isotopically labeled proteins. The secondary structure of the protein was deduced from distance measurements in the micelles and the arrangement of these secondary structures within the lipid matrix deduced from angular measurements with oriented layers. The dynamics of the protein were determined from motional narrowing of line shapes in the solid-state experiments.

The distance information on this protein in solution was obtained from ^{1}H-^{1}H homonuclear nuclear Overhauser effect (NOE) measurements and used to determine secondary and tertiary structure. With the bacteriophage, protein labeling took place with ^{2}H (80%) on all carbon sites, with ^{15}N (98%) on all nitrogen sites, and with ^{2}H (50%) on all exchangeable sites, and by combining ^{1}H-^{15}N heteronuclear correlation and ^{1}H-^{1}H NOE spectroscopy. This led to the conclusion that this protein contained two helical segments. One of these could fit into the lipid matrix while the other was in the water (Shon et al., 1991).

Another spectroscopic technique that has been applied to the study of biomembrane systems is Fourier transform infrared (FTIR) spectroscopy. This technique has become quite popular since technical developments have made it possible to obtain good infrared spectra of biomolecules in H_2O. Previously the strong absorption of water had made it difficult to obtain information about the protein structure, due to overlap of the water absorption with that of the protein.

Experiments have been made with model membranes consisting of lipid-water systems and also reconstituted systems containing polypeptides or proteins as well as natural biomembranes. The phase transitions of the lipid-water systems are readily studied using FTIR, and the effect of cholesterol or poly-peptides on the phase transition have been examined. The results are in agreement with those obtained using techniques such as calorimetry or NMR spectroscopy. Studies of membrane proteins have also been carried out and qualitative and quantitative studies have been made (Haris and Chapman, 1992).

The major band in the FTIR spectra of proteins is the amide I band. This absorption arises predominantly from the $C=O$ stretch of the peptide bonds and as such is sensitive to the hydrogen bonding state of the protein. The different secondary structures present in a protein are each associated with a characteristic hydrogen bonding pattern, and so are each associated with a characteristic amide I frequency. The presence of a range of secondary structures in membrane proteins results in the production of multiple amide absorptions. The half-width of each absorption is such that they cannot be resolved instrumentally and a composite band results. Recently developed mathematical resolution enhancement techniques, such as second derivative and deconvolution analysis, have been used to detect the presence of individual components beneath the broad amide I contour of a number of proteins. These enhancement techniques, coupled with hydrogen-deuterium exchange studies, have been used to detect small changes in protein structure.

FTIR spectra of a large number of membrane proteins obtained in both H_2O and ^2H_2O reveal that the proteins are predominantly α-helical in structure (Haris and Chapman, 1992). Furthermore, the amide I maxima for the α-helical structure in membrane protein spectra differ from the corresponding maxima in the spectra of soluble proteins (Haris et al., 1989).

7.9.1. Bacteriorhodopsin and Rhodopsin

Bacteriorhodopsin and rhodopsin are two light-transducing transmembrane proteins.

Bacteriorhodopsin is found in the purple membrane of *Halobacterium halobium*. The protein contains a chromophore, all-*trans* retinal, which isomerizes to 13-*cis* retinal upon absorption of a photon. The pioneering work of Henderson and Unwin (1975) using electron diffraction and electron microscopy has contributed greatly to our knowledge of the structure of bacteriorhodopsin. It has been suggested that rhodopsin has a structure in the membrane similar to that shown for bacteriorhodopsin. The supposed structural similarities are striking. For example, the two proteins occupy roughly the same cross-sectional area in the membrane, and the retinal attachment lysine is located in approximately the same vertical position on the C-terminal transmembrane helix.

The comparison between the infrared spectra of rhodopsin and bacteriorhodopsin, in both H_2O (see Fig. 7.7) and ^2H_2O, reveals marked similarities and some differences. The amide I band maximum observed with rhodopsin $(1,657 \text{ cm}^{-1})$ is similar to that observed with a number of membrane proteins such as the ATPases (Na^+/K^+, H^+/K^+, Ca^{2+}), glucose transporter, cytochrome c oxidase, and bacterial and higher plant reaction centers (Haris et al.,

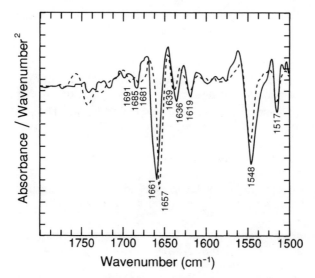

Fig. 7.7. Comparison of the second derivative infrared spectra of bacteriorhodopsin (———) and rhodopsin (---) obtained in H_2O. From Haris et al. (1989), with permission of the publisher.

1989). The high-frequency amide I band of bacteriorhodopsin is unusual. Some workers suggest that it is due to the presence of α_{II}-helices. An alternative explanation for the high-frequency amide I band is the possible existence of short 3_{10}-helical regions in addition to the normal α-helical structure.

REFERENCES AND FURTHER READING

Bar, R.S., Deamer, D.W. and Cornwell, D.G. (1966) Surface area of human erythrocyte lipids: Reinvestigation of experiments with plasma membranes. Science *153*, 1010–1012.

Bretscher, M.S. (1972) Phosphatidylethanolamine: Differential labelling in intact cells and cell ghosts of human erythrocytes by a membrane-impermeable reagent. J. Mol. Biol. *71*, 523–528.

Chapman, D. (1984) Biological Membranes, Vol. 5., D. Chapman, Ed. Academic Press, United Kingdom.

Chapman, D., Byrne, P. and Shipley, G.G. (1966) The physical properties of phospholipids. Part 1: Solid state and mesomorphic properties of some 2,3-diacyl-DL-phosphatidylethanolamines. Proc. Roy. Soc. Ser. A. *290*, 115–142.

Cowan, S.W. and Rosenbusch, J.P. (1994) Folding pattern diversity of integral membrane proteins. Science *264*, 914–916.

Danielli, J.F. and Davson, H. (1935) A contribution to the theory of permeability of thin films. J. Cell Comp. Physiol. *5*, 483–494.

Davies, A., Ciardelli, T.L., Lienhard, E., Boyle, J.M., Whetton, A.D. and Baldwin,

S.A. (1990) Site-specific antibodies as probes of the topology and function of the human erythrocyte glucose transporter. Biochem. J. *266*, 799–808.

Devaux, P.F. (1991) Static and dynamic lipid asymmetry in cell membranes. Biochemistry *30*, 1163–1173.

Dohlman, H.G., Caron, M.G. and Lefkowitz, R.J. (1987) A family of receptors coupled to guanine nucleotide regulatory proteins. Biochemistry *26*, 2657–2664.

Dohlman, H.G., Thorner, J., Caron, M.G. and Lefkowitz, R.J. (1991) Model systems for the study of seven transmembrane segmant receptors. Annu. Rev. Biochem. *60*, 653–664.

Gorter, E. and Grendel, F. (1925) Bimolecular layers of lipids on chromatocytes of blood. J. Exp. Med. *41*, 439–443.

Haris, P.I. and Chapman, D. (1992) Does Fourier transform infrared spectroscopy provide useful information on protein structures? Trends Biochem. Sci. *17*, 328–333.

Haris, P.I., Coke, M. and Chapman, D. (1989) Fourier transform infrared spectroscopic investigation of rhodopsin structure and its comparison with bacteriorhodopsin. Biochim. Biophys. Acta *995*, 160–167.

Henderson, R. (1977) The purple membrane from *Halobacterium halobium*. Annu. Rev. Biophys. *6*, 87–109.

Henderson, R. and Unwin, P.N.T. (1975) Three-dimensional model of purple membrane obtained by electron microscopy. Nature *257*, 28–30.

Kamp, J.A.F. Op den, (1979) Lipid asymmetry in membranes. Annu. Rev. Biochem. *48*, 47–71.

Luzzati, V., Gulik, A., Gulik-Krzywicki, T. and Tardieu, A (1986) Lipids and membranes: Past, present, and future. J.A.F. Op den Kamp, B. Roelofsen, and K.W. Wirtz, Eds. Elsevier, Amsterdam, p. 137.

Mackinnon, R. and Miller, C. (1989) Mutant potassium channels with altered binding of charybdotoxin, a pore blocking peptide inhibitor. Science *245*, 1382–1385.

Miller, C. (1992) Hunting for the pore of voltage gated channels. Curr. Biol. *2*, 573–575.

Quinn, P.J. and Chapman, D. (1991) in Fundamentals of Medical Cell Biology, Vol. 2: Structural Biology. E. Edward Bittar, Ed. Jai Press, Greenwich, Conn.

Shon, K.J., Kim, Y. Colnago, L.M. and Opella, S.J. (1991) NMR studies of the structure and dynamics of membrane-bound bacteriophage. Science *252*, 1303–1305.

Singer, S.J. and Nicolson, G.L. (1972) The fluid mosaic model of the structure of cell membranes. Science *175*, 720–731.

von Heijne, G. (1988) Transcending the impenetrable: How proteins come to terms with membranes. Biochim. Biophys. Acta *947*, 307–333.

Wieslander, Å., Christiansson, A., Rilfors, L. and Lindblom, G. (1980) Lipid bilayer stability in membranes, Regulation of lipid composition in *Acholeplasma laidlawii* as governed by molecular shape. Biochemistry *19*, 3650–3655.

CHAPTER 8

RECONSTITUTION OF MEMBRANE FUNCTION IN BILAYER SYSTEMS

8.1. THE OBJECTIVES OF RECONSTITUTION

The complexity of biological membranes in terms of their lipid, protein, and glycoprotein composition and the multitudinous nature of their functions leads naturally to the desire to separate their various components so they can be studied in simpler systems and possibly in isolation. There are clearly drawbacks to this philosophy, in that many membrane processes are dependent on linked sequences of events involving several membrane proteins functioning in a coordinated manner; however, by learning something about the individual components it is possible to make reasonable deductions about the sequence of events which take place *in vivo* and so build up theories of the mechanism of processes which can be further tested on isolated cells. The advent of liposomes and planar bilayer lipid membranes (BLMs) in the last 30 years has enabled many membrane proteins such as transporters and receptors to be reassessed in these model bilayer systems, and has led to the development of reconstitution studies in which membrane components are successively isolated from natural membranes and then put back into model membrane systems, where their characteristics can be more easily investigated.

The objectives of reconstitution studies relate to a variety of membrane properties, among which the following may be obtained:

1. Information on the structure of membrane proteins/glycoproteins in the lipid bilayer assembly, such as the secondary structure of transmembrane elements of integral proteins and the disposition of polypeptide chains in

the bilayer (i.e., the number of times a polypeptide chain crosses the bilayer) and the arrangements of subunits in the bilayer

2. Information on the function of membrane proteins, particularly membrane receptor and transporter molecules, such as the binding of ligands to receptors, particularly hormones, and the transport of ions and metabolites across the bilayer

3. Information on how structure and function of membrane proteins/glycoprotein are influenced by the composition and physical state of the bilayer lipid environment

4. Information on the contributions particular membrane proteins/glycoproteins make to the surface properties of cell membranes and how such properties affect the nature of interactions between cells

The anchoring of membrane proteins and glycoproteins to the bilayer of cell membranes depends largely on the hydrophobic interaction between bilayer lipids and the transmembrane sequences of the integral membrane proteins—or hydrophobic and/or ionic interactions in the case of extrinsic membrane proteins. Thus reconstitution experiments depend initially on the breakdown of some or all of these interactions in a manner that causes as little change as possible in the conformation of the membrane proteins and hence with the retention of function. This can generally be achieved with the aid of a nondenaturing detergent to break down the membrane and solubilize the membrane proteins/glycoproteins. After separation of the protein required to be reconstituted, the hydrophobic interaction must be exploited to incorporate it into the bilayer of the model membrane system. Several types of model system can be used: (1) liposomes (vesicles), (2) planar bilayer lipid membranes (BLMs), and (3) vesicles prepared by fusion of phospholipid liposomes with isolated biomembrane vesicles of cellular origin.

Clearly, with the diversity of membrane proteins, it is not possible to define a universal method of reconstitution that will work for all proteins. Thus a number of strategies have been developed that work for individual proteins (Jones et al., 1987b). In this chapter some of these are discussed, with particular reference to receptors and transporters.

8.2. GENERAL PRINCIPLES OF RECONSTITUTION

The general principles relating to biological membrane solubilization were discussed in Chapter 6 (sec. 6.1–6.3). We consider here the methods used to reverse the solubilization process—to reassemble membrane components in bilayer systems, most frequently liposomes or vesicles. The solubilization of membranes generally results in the formation of mixed micelles of lipid and surfactant, together with protein-surfactant and/or protein-lipid-surfactant complexes. In such systems, the ratios of surfactant to lipid and protein are high. Reconstitution involves the complete or partial removal of surfactant in such a way that the lipid is returned to the bilayer state with the membrane protein inserted in it, preferably in its native conformation.

There are several approaches to achieving the final objective, the overall strategy being to progressively reduce the surfactant-lipid ratio in the solubilized system such that vesicles are formed incorporating the protein. Dialysis or gel filtration are the most convenient ways of removing surfactant. Figure 8.1 shows possible mechanisms for reconstitution from the surfactant solubilized state. In Figure 8.1A, surfactant removal from a mixture of surfactant-lipid mixed micelles and protein-lipid-surfactant complexes results in the reformation of a bilayer phase, which on further surfactant removal forms sealed vesicles. The orientation of the membrane protein in the vesicles will most likely be symmetric, since the protein has an equal probability of being distributed on either side of the bilayer, although the size of the vesicles and hence radius of curvature may influence the symmetry. In contrast, if solubilized surfactant-protein complexes are added to preformed vesicles and the surfactant is gradually removed, the protein will insert into the vesicle bilayer. In this case, the

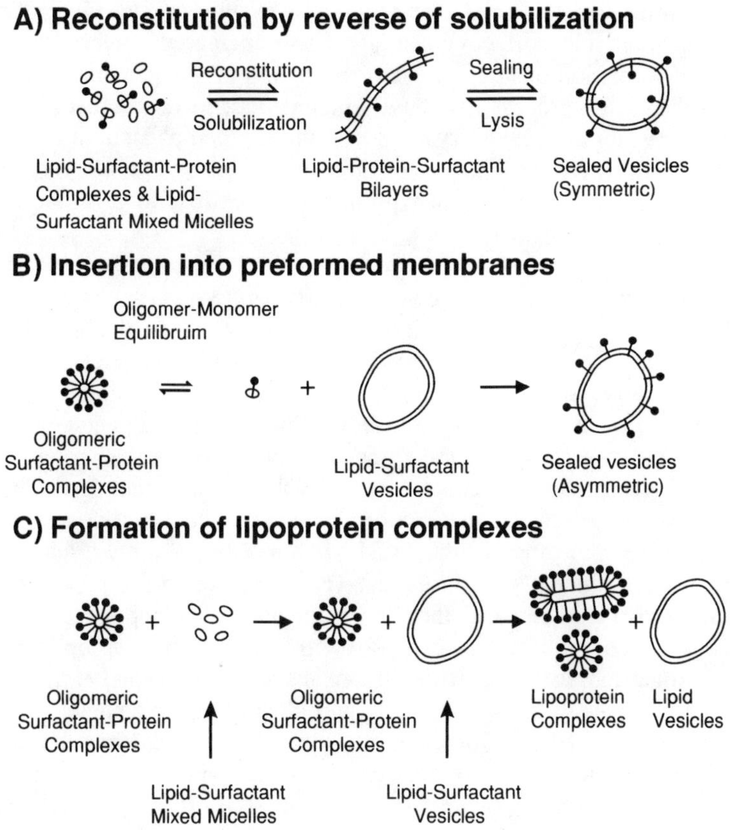

A) Reconstitution by reverse of solubilization

Lipid-Surfactant-Protein
Complexes & Lipid-
Surfactant Mixed Micelles

Lipid-Protein-Surfactant
Bilayers

Sealed Vesicles
(Symmetric)

B) Insertion into preformed membranes

Oligomeric
Surfactant-Protein
Complexes

Lipid-Surfactant
Vesicles

Sealed vesicles
(Asymmetric)

C) Formation of lipoprotein complexes

Oligomeric
Surfactant-Protein
Complexes

Oligomeric
Surfactant-Protein
Complexes

Lipoprotein
Complexes

Lipid
Vesicles

Lipid-Surfactant
Mixed Micelles

Lipid-Surfactant
Vesicles

Fig. 8.1. Suggested mechanisms for the reconstitution of membrane proteins into lipid vesicles. From Helenius et al. (1981).

orientation will most probably be asymmetric (Fig. 8.1B). A further mechanism may involve the initial formation of protein-free vesicles from surfactant-lipid mixed micelles, followed by the insertion of protein from protein-surfactant complexes as the surfactant-lipid ratio is reduced during dialysis (Fig. 8.1C). This latter mechanism would occur if the protein-surfactant complexes were more stable than the surfactant-lipid mixed micelles, so that the rate of removal of surfactant was greater from the mixed micelles than from the protein complexes. Whether reconstitution occurs by mechanism A or C when surfactant is dialysed out will of course depend on the protein, the lipid present, and the surfactant used.

A modification of insertion method B that has been widely used involves the introduction of a freeze-thaw-sonication step. Here the solubilized protein extract is mixed with vesicles and the mixture rapidly frozen in an acetone or methanol dry ice mixture (or in liquid nitrogen), thawed to room temperature, and briefly sonicated (~ 30 sec). The freeze-thaw-sonication procedure was first introduced by Kasahara and Hinkle (1977) for reconstituting the glucose transporter from the human erythrocyte into small unilamellar liposomes. The procedure improved the uptake of D-glucose by the vesicles, indicating that the efficiency of transporter reconstitution was increased. Freeze-thawing results in the formation of larger vesicles by fusion of smaller ones. The mechanism of the fusion process is not fully understood; it has been suggested that on freezing, water molecules crystallize on the charged phospholipid interfaces and form planes between bilayers of adjacent vesicles. The bilayers are then casily fractured and the exposed hydrophobic cores fuse to give rise to larger vesicles during slow thawing (Pick, 1981).

The most commonly used reconstituted systems are composed of liposomes or vesicles; planar bilayer lipid membranes (BLMs) supported across an orifice in a hydrophobic sheet (Teflon) have also been used, particularly for sugar transport studies (Jones and Nickson, 1981). BLMs have the advantage of well-defined geometry, and transport between the two aqueous compartments on either side of the BLM is simple to follow. The major disadvantage, however, is that it is difficult to control the extent of penetration of the protein into the bilayer and even more difficult to measure the amount that has penetrated. Thus, in contrast to vesicles that can be separated by gel filtration and analyzed in terms of the lipid-protein ratio, a BLM cannot be as simply characterized. BLMs are also inherently unstable and very sensitive to the presence of free surfactant. For these reasons, they have not proved as popular as liposomes for reconstitution studies. However, they are particularly appropriate for the study of single ion channels, where the existence of a channel can be directly monitored electrically and the opening and closing of the channel can be followed (Coronado, 1986). For example, vesicles derived from the epithelial cells of the colonic mucosa of rats can be fused with BLMs to reconstitute single-anion channels. Spontaneous switching (on-off) of the reconstituted channels can be observed for a series of halide anions when a voltage of 30 mV is applied across the bilayer in the presence of halide ion concentration of 200 mM (Reinhardt et al., 1987).

8.3. CHOICE OF SURFACTANT FOR RECONSTITUTION STUDIES

As discussed in Chapter 1, there is a very wide range of surfactants from which to choose for a particular solubilization and reconstitution study. Given such a choice, it is important to have some guidelines to making a satisfactory decision for a particular reconstitution. In practice, despite the wide range of surfactants available, a relatively limited number have actually been used. This situation has arisen because research workers have tended to keep to surfactants that have been successfully used in previous studies, an approach that inevitably leads to the use of a restricted range of materials. The choice of surfactant is based on several important considerations, of which the following factors are generally significant.

1. The surfactant must readily solubilize the protein/glycoprotein from its natural environment (e.g., cell membrane) without causing denaturation. Thus the choice of a nondenaturing surfactant is generally of paramount importance. Nonionic surfactants such as the Tritons, the Lubrols, and octyl β-D-glucopyranoside (OBG) are commonly chosen. Of the anionics, the cholates (sodium cholate and deoxycholate) are also favored. Surfactants such as the n-alkyl sulphates are not generally suitable, since while they very effectively solubilize they are powerful denaturants. In exceptional circumstances, denaturation does not occur, e.g., sodium n-dodecylsulphate does not denature bacterial catalase (from *Micrococcus luteus*) (Jones et al., 1982) and in fact activates *Aspergillus niger* catalase (Jones et al., 1987a). There are also instances of enzyme activation by Triton X-100 (which activates glucose-6-phosphatase [Beyhl, 1986]) and deoxycholate (which activates phospholipase C [El-Sayert and Roberts, 1985]), but such effects are rather uncommon, since they depend on the surfactant slightly changing the active site of an enzyme or binding site of a receptor or transporter to a slightly more favorable conformation than it was in the native state, which is unusual.

2. Once solubilized, the surfactant must be progressively removed to effect a reconstitution as described above. Since dialysis is commonly used to do this and the rate of dialysis depends on the diffusion of "monomeric" surfactant— as distinct from solubilized complexes or mixed micelles—the concentration of monomeric surfactant in equilibrium with the complexes will determine the rate of dialysis. The concentration of free monomer will depend on the critical micelle concentration (cmc) in the medium used. Surfactants with large cmcs are thus preferred, because the monomer concentration gradient across the dialysis membrane will be high and the surfactant more rapidly removed. In this respect, the Tritons and Lubrols, having very low cmcs, are not ideal, and for this reason many workers have favored OBG, which has an unusually high cmc (~ 23 mM) and is rapidly dialysed out.

3. During a reconstitution study, it is important to be able to follow the progress of the solubilization and reconstitution by using an appropriate assay for the protein/glycoprotein of interest and the ratio of protein/glycoprotein to lipid in the system. This requires an appropriate assay that is not interfered

with by the surfactants. It is important to screen possible surfactants for their effects on any assays that might be used during the reconstitution. For example, the phenol residue in the Tritons absorbs strongly at 280 nm (the absorption maximum of proteins), and hence protein absorption at 280 nm cannot be used to assay for protein in Triton solutions. To circumvent this problem, reduced Tritons can be used in which the phenyl group is reduced to cyclohexyl (Tiller et al., 1984). The cmc of reduced Triton X-100 is approximately 12% higher than that of the dehydrogenated form, which is an added advantage.

4. The stability of the surfactant is also important, especially in reconstitution experiments that take several days to complete. In this respect, unsaturated surfactants or those prone to hydrolysis and/or autoxidation should be avoided.

There are a variety of other considerations that might be important in choice of surfactant, including, for example, ease of purification, toxicity, performance on ion-exchange chromatography, and availability in a radiolabeled form (Jones et al., 1987b). Not the least of considerations is the cost, especially when large amounts are required. In this respect, OBG is an expensive material to use on a large scale; however, n-octyl β-D-thioglucopyranoside is cheaper to produce and is more stable to hydrolysis (Saito and Tsuchiya, 1984). The cmc of the thioglucoside is 9 mM (cf. 23 mM for OBG) but still high enough to be easily dialysed out.

8.4. RECONSTITUTED MEMBRANE PROTEIN/GLYCOPROTEIN SYSTEMS

A large number of membrane proteins and glycoproteins have now been reconstituted in model membranes, particularly in phospholipid vesicles or liposomes of various types. Tables 8.1 to 8.3 show some examples of transporters (Table 8.1), receptors (Table 8.2), and other proteins and glycoproteins (Table 8.3) that have enzymic or structural functions in membranes. An extensive list of proteins that have been reconstituted in bilayers has been given by Jain and Zakin (1987), and Levitzki (1985) describes a number of receptor reconstitutions. Reconstituted systems offer a means of investigating factors that influence the function of the membrane constituents *in vitro*. In the case of intrinsic membrane proteins, which include the transporters and many receptors, a major proportion of the protein is in intimate contact with the lipid of the bilayer. The protein is in effect a solute in a lipid solvent, although the "solution" is restricted to two dimensions and the fluidity is much less than that of a typical liquid. The integral membrane protein cannot flip-flop from one side of the bilayer to the other, but is restricted to diffusion laterally within the plane of the bilayer. The lateral diffusion coefficient of an integral membrane protein is of the order of 10^{-10} cm^2 s^{-1}, while the lipids have a lateral diffusion coefficient of 10^{-8} cm^2 s^{-1}. In contrast to the proteins, the lipids can

TABLE 8.1. Some Reconstitution Studies on Membrane Transporters

Transporter	Function	Solubilizer	Reconstituted System
H[+]-ATPase[a]	ATP synthesis	Triton X-100	Reverse phase evaporation (large unilamellar liposomes) egg PC + PA
Ca[2+]-ATPase[b]	Ca[2+] pump of sarcoplasmic reticulum	Triton X-100	Reverse phase evaporation (large unilamellar liposomes) egg PC + PA
Bacteriorhodopsin[c]	H[+] pump of *Halobacterium Halobium*	Triton X-100	Reverse phase evaporation (large unilamellar liposomes) egg PC + PA
Bacteriorhodopsin[d]	H[+] pump of *Halobacterium Halobium*	None—sonicated aggregates of bacteriorhodopsin	Sonicated unilamellar vesicles (DMPC and DOPC)
Lactose[e]	Lactose translocation	Octyl glucoside	Unilamellar vesicles of numerous phospholipids (freeze-thaw-sonication)
Erythrocyte anion transporter[f]	Cl[-]/HCO$_3^-$ exchange	Octaoxyethylene mono-n-dodecyl ether (C$_{12}$E$_8$)	Multilamellar vesicles of numerous phospholipids
Erythrocyte D-glucose transporter[g]	D-glucose transport	Triton X-100	Large unilamellar vesicles of numerous phospholipids
Adipocyte D-glucose transporter[h]	D-glucose transport	Sodium cholate	DOPC/cholesterol one transporter per vesicle
Bovine heart glucose[i] transporter	D-glucose transport	Sodium cholate	Soybean phosphatidylcholine (freeze-thaw-sonication)

Data from: [a]Lévy et al. (1990); [b]Lévy et al. (1990); [c]Lévy et al. (1990); [d]Scotto and Zakin (1988); [e]Garcia et al. (1983); [f]Maneri and Low (1988); [g]Tefft et al. (1986); [h]Gorga and Lienhard (1984); [i]Wheeler and Hauck (1985).

TABLE 8.2. Some Reconstitution Studies on Membrane Receptors

Receptor	Ligands	Solubilization	Reconstituted System
Acetylcholine (from *Torpedo californica*)[a]	Acetylcholine, α-bungarotoxin	Sodium cholate	Asolection and phospholipid vesicles
β-Adrenergic receptor (rabbit liver)[b]	G_s protein and adenylate cyclase	Lubrol-PX	Phospholipid vesicles containing turkey erythrocyte β-adrenergic receptors
Immunoglobulin E (from rat ascitic fluid)[c]	IgE	Triton X-100, CHAPS, sodium cholate, octylglucoside	Egg and soybean lecithin vesicles
Asialoglycoprotein receptor (from rabbit hepatocytes)[d]	Desialylated serum glycoprotein	Detergent-free system	Sonicated DPPC vesicles
Insulin (rat adipocyte)[e]	Insulin	Sodium cholate	Egg lecithin and other phospholipid vesicles (freeze-thaw-sonication)

Data from: [a]Jones et al. (1987b); [b]Pederson and Ross (1982); [c]Rivnay and Metzger (1982); [d]Klausner et al. (1980); [e]More and Jones (1983), Jones et al. (1986).

TABLE 8.3. Reconstitution of Membrane Proteins/Glycoproteins

Protein/Glycoprotein	Function	Solubilizer	Reconstituted System
Rhodopsin[a]	Visual pigment of retina	Octyl glucoside	Various phospholipid vesicles produced by dialysis
Spike glycoprotein (Semliki forest virus)[b]	Virus coat	Octyl glucoside	Egg lecithin vesicles
Penicillinase (Bacillus licheniformis membrane)[c]	Amide bond cleavage of penicillin β-lactam ring	Octyl glucoside	Egg lecithin vesicles
Sucrose Isomaltase (rabbit small intestine brush border)[d]	Sucrose hydrolysis	Triton X-100/sodium cholate	Egg lecithin unilamellar vesicles
Guanylate cyclase (from sea urchin spermatozoa)[e]	GMP metabolism	Lubrol PX	Various phospholipids, small unilamellar vesicles
Acyl carnitine (from rat liver)[f]	Fatty acid metabolism and acyl group transport	Octyl glucoside	Sonicated asolectin vesicles
M13 bacteriophage coat protein[g]	Bacteriophage structure	Sodium cholate	PS/PC multilamellar vesicles (freeze-thaw-sonication)
UDP glucuronosyltransferase (from pig liver microsomes)[h]		Triton X-100	DMPC/cholesterol vesicles
Cytochrome oxidase (from beef heart mitochondria)[i]	Mitochondrial electron-transport chain enzyme	Triton X-100	DMPC/cholesterol vesicles
Glycophorin (from human erythrocyte membrane)[j]	Blood group carrier, imparts negative charge to cell surface	Phenol/water partition	DMPC/DPPC/DSPC sonicated vesicles
Chlorophyll a/b light harvesting complex (from peas)[k]	Photosynthetic system	Triton X-100	Soybean lecithin vesicles (freeze-thaw-sonication)

Data from: [a]Jackson and Litman (1982); [b]Helenius et al. (1981); [c]Helenius et al. (1981); [d]Brunner et al. (1978); [e]Radany et al. (1985); [f]Noel et al. (1985); [g]Florine and Feigenson (1987); [h]Scotto and Zakim (1986); [i]Scotto and Zakim (1986); [j]Goodwin et al. (1982); [k]McDonnel and Staehelin (1980).

flip-flop, although in general at a rate much lower than that of lateral diffusion. The immediate lipid environment of an intrinsic membrane protein will thus be continuously changing, although some of the membrane lipids might have, on average, a preference for interaction with the hydrophobic transmembrane domains of a particular integral protein. It is appropriate to consider the influence the lipid environment has in membrane protein function.

8.4.1. The Role of Lipid in Membrane Transporter Function

The physical slate of membrane lipids is markedly influenced by temperature, the most dramatic change occurring when lipids pass through the gel to liquid-crystalline phase transition ($L_{\beta^1} \rightarrow L_{\alpha}$). This transition, which is highly cooperative, occurs when the acyl chains in the all-*trans* conformation in the gel state "melt" with the introduction of *trans*-gauche conformation changes (see Chap. 7). Such a change in physical state would be expected to influence the properties of membrane-associated proteins; however, in natural membranes the effects on transport, enzyme activity, or ligand binding to a receptor cannot be separated from any concomitant changes in lipid composition in the environment of the protein. However, the reconstituted system is ideal for investigating the effects of bilayer fluidity and lipid composition on membrane proteins, in that the lipid environment can be controlled and the system can be investigated through the lipid chain-melting temperature (T_m). In the case of transport proteins, the glucose and anion transporters of the human erythrocyte membrane are two of the most extensively studied passive transport systems (i.e., transport occurs down a concentration gradient by facilitated diffusion and requires no metabolic energy). The structures of these transporters have been discussed (see Chap. 1); both are single chains with the glucose transporter (relative molecular mass 55,000) believed to have 12 α-helical membrane-spanning domains and the anion transporter (relative molecular mass 90,000) 14 α-helical membrane-spanning domains.

The effects of lipid environment on the glucose transporter have been studied in some detail (Tefft et al. 1986; Carruthers and Melchior, 1986) for the transporter reconstituted in large unilamellar vesicles prepared by reverse-phase evaporation from a range of different phospholipids. The activation energies for passive sugar transport into vesicles of phospholipids having the same acyl chains but different headgroups differ depending on the headgroup, as shown by the data in Table 8.4. The reconstituted vesicles are not identical in terms of size and the number of transporter molecules per vesicle, as reflected by the number of cytochalasin B (a sugar transport inhibitor) binding sites. However, the activation energies for transport—particularly in the case of the egg phospholipids—depend markedly on headgroup. Figure 8.2 shows the dependence of activation energy on acyl chain length and the state of the bilayer for phosphatidylcholines (PCs) and phosphatidylglycerols (PGs). For PGs the chain-length dependence of activation energy differs in the gel and liquid-crystalline states, whereas for the PCs the dependence is similar in both states.

TABLE 8.4. Physical Characteristics of the Human Erythrocyte D-Glucose Transporter in Reconstituted Vesicles

Lipid	Vesicle Diameter (nm)	CCB Binding[a] Sites per Vesicle	P[b] ($\times 10^6$ cm s^{-1}) at 50°C	E_a[c] (kJmol^{-1} K^{-1}) (Temp. °C)
DMPG	170	1	0.34 ± 0.04	1.82 (24°)
DMPG	170	1	0.34 ± 0.04	58.2 (50°C)
DMPA	480	33	2.6 ± 0.3	16.4 (24°C)
DMPS	1090	66	2.75 ± 0.4	69.1 (50°)
Egg PC	690	17	0.79 ± 0.05	33.7 (50°)
Egg PG	480	10	2.14 ± 0.3	50.0 (50°)
Egg PA	540	8	3.1 ± 0.2	18.2 (50°)
Egg SM	562	1152	0.98 ± 0.06	58.2 (24°)

Data from Tefft et al. (1986).
[a]CCB is cytocholasin B a transport inhibitor.
[b]Permeability coefficient.
[c]Activation energy.

The kinetics of facilitated diffusion can be broadly described in terms of Michaelis-Menten kinetics and the kinetic parameters (K_m and V_{max}) measured through the gel to liquid-crystalline phase transition. Figure 8.3 show K_m and V_{max} as a function of temperature for the sugar transporter reconstituted in DMPC and DMPG vesicles. DMPC does not support transport below the phase transition—transport coincides with the onset of chain-melting. In contrast,

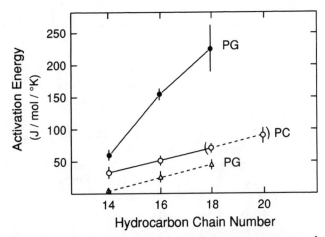

Fig. 8.2. Dependence of activation energy for D-glucose transport on acyl chain length in vesicles incorporating the human erythrocyte sugar transporter. The solid lines are for vesicles in the liquid-crystalline state and the dashed lines for the gel state. From Tefft et al. (1986), with permission of the American Chemical Society.

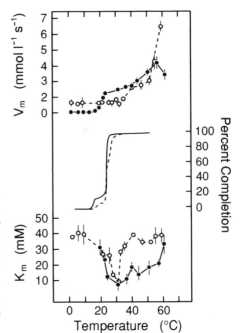

Fig. 8.3. Dependence of the Michaelis-Menten kinetic parameter on temperature for the human erythrocyte sugar transporter reconstituted in DMPC (solid line) and DMPG (dashed line) vesicles. The percentage completion of the transition from the gel to the liquid-crystalline state is shown in center of the figure. From Tefft et al. (1986), with permission of the American Chemical Society.

V_{max} for DMPG remains constant through the transition but increases exponentially between 40°–60°C (the transporter is thermally denatured at 60°–62°C). For both lipids, K_m decreases on chain melting, suggesting a stronger interaction between transporter and sugar in the region of coexistence of the gel and liquid-crystalline phase; above T_m, K_m increases more steeply for DMPG than for DMPC. These studies lead to the conclusion that both the lipid headgroup and the acyl chain length have significant effects on transport activity.

Studies on the anion transporter using a different experimental approach also lead to a similar conclusion. Maneri and Low (1988) investigated the anion transporter reconstituted in a range of phospholipid vesicles by differential scanning calorimetry. The experiments gave information on the structural stability of the integral domain of the transporter in terms of the temperature of maximum heat capacity (i.e., the denaturation temperature, T_m) and the enthalpy of denaturation. In order to focus attention on the 55,000 molecular mass membrane-spanning domain of the transporter, the 43,000 molecular mass cytoplasmic domain was removed by proteolysis. For a given phospholipid headgroup, the value of T_m increases markedly with acyl chain length for monounsaturated symmetric phosphatidylcholines, as shown in Figure 8.4. The change in T_m from the protein reconstituted in dimyristoleylphosphatidylcholine vesicles (C14 : 1) to the dinervonylphosphatidylcholine vesicles (C24 : 1) is substantial (47° to 66°C), demonstrating that the stability of the transmembrane domain of the transporter is very dependent on the nature of the surrounding lipid. It is significant that in the native erythrocyte membrane T_m is 68°C,

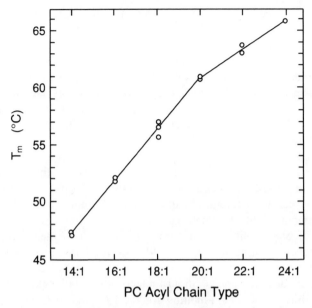

Fig. 8.4. The effect of acyl chain length on the temperature of maximum heat capacity (T_m, denaturation temperature) for the membrane-spanning domain of the human erythrocyte anion transporter reconstituted in monounsaturated phosphatidylcholine vesicles. From Maneri and Low (1988), with permission of the publisher.

higher than in any of the reconstituted systems, so that not all the factors that are essential for stabilization of the transporter in its native state are present even in the vesicles prepared from the longest acyl chain phosphatidylcholine.

Increasing the degree of saturation in a series of diacyl vesicles (C18:1, C18:2, C18:3) decreased T_m in the series 56°C to 53.5°C to 49°C, respectively. Changing the lipid headgroup of the C18:1 phosphatidyl moiety gave T_m values of 38°C (phosphatidylglycerol), 47°C (phosphatidylserine), 56°C (phosphatidylethanolamine), and 56°C (phosphatidylcholine). For dioleyl-phosphatidyl-choline vesicles, addition of cholesterol increased the T_m linearly from 55.5°C (O mole % cholesterol) to 62°C (43 mole % cholesterol).

It is clear from these observations that the acyl chain length and degree of unsaturation have a considerable effect on the thermal stability of the membrane-spanning domain of the transporter. Given that the protein has a well-defined hydrophobic cross-section, then optimal stability will occur when the hydrophobic transmembrane domain of the protein closely matches the nonpolar zone of the bilayer lipid. That is, there must be hydrophobic complementarity between protein and bilayer. The fact that stability increases with increasing acyl chain length suggests that the transmembrane domain of the transporter is longer than the C24:1 acyl chain. The hydropathy plot of the transporter indicates that there are 28 amino acid residues in each of the putative 12 membrane-spanning α-helices, which suggests a hydrophobic zone with a width of

4.2 nm, which exceeds the width of the average fluid membrane ($\sim$3 nm) by a considerable margin. Stability would thus be greater in longer acyl chain lipids in the gel (all-*trans*) states. Introducing unsaturation results in a reduced bilayer thickness and hence reduced hydrophobic complementarity, while addition of cholesterol thickens the bilayer with a resulting increase in complementarity. The effects of different headgroups is probably largely an electrostatic effect. Thus PE and PC, which are zwitterionic, behave similarly with approximately the same T_m, whereas the negatively charged PG and PS destabilize the transporter.

The concept of hydrophobic complementarity is also important in the case of vesicles incorporating glycophorins from the human erythrocyte membrane (Goodwin et al., 1982). Glycophorin A has a molecular mass of approximately 31,000 and spans the membrane with a single 23 amino acid residue α-helix. When incorporated into phosphatidylcholine unilamellar vesicles, it results in withdrawal of phospholipid from participation in chain melting, so that the enthalpy of chain melting is reduced, although the chain-melting temperature is not markedly changed. When reconstituted in DMPC, DPPC, and DSPC vesicles, the numbers of phospholipid molecules per molecule of glycoprotein that are withdrawn from chain melting are 42, 197, and 240, respectively. The α-helix disrupts the lipid packing, and the extent of disruption increases with acyl chain length. In this case, the hydrophobic membrane-spanning domain is 3.45 nm long, whereas the bilayer thicknesses for DMPC, DPPC, and DSPC in the all-*trans* conformation are 4.0, 4.64, and 5.20 nm, respectively. Thus hydrophobic complementarity is better for the shorter acyl chain phospholipid, whereas to obtain complementarity with long-chain lipids requires the introduction of *trans*-gauche kinks leading to greater lipid disorganization around the α-helical chain. The numbers of concentric "shells" of disorganized (melted) lipid increase in the series 4, 9, and 10 for DMPC, DPPC, and DSPC, respectively.

It is clear from the above discussion that the choice of phospholipid for transporter reconstitution can critically affect the behavior of the transporter in the reconstituted state. It follows that it will not generally be satisfactory to compare the properties of a reconstituted system with those of the native membrane without taking into account the possible modulating effects of the lipid used.

8.4.2. The Role of Lipid in Membrane Receptor Function

The effect of lipid environment on receptor function in reconstituted systems is not well understood, largely because it has not been studied to any great extent (Levitzki, 1985). In contrast to membrane transporters, the functional domains of receptors are located on the outer surface of the bilayer in a peripheral position. The membrane-penetrating hydrophobic domains serve to anchor the receptor but have relatively little influence on the peripheral binding domain. For example, the insulin receptor has an $\alpha_2\beta_2$ structure (Fig. 8.5) with a total molecular mass of approximately 350,000. The insulin binding site

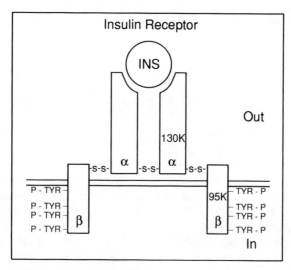

Fig. 8.5. The proposed structure of the insulin receptor showing the disulphide linkages between the subunits and the phosphorylation sites in the β subunit cytoplasmic domains. From Carpentier (1989), with permission of the publisher and author.

is located in the α_2 peripheral domain and the receptor is anchored to the bilayer by the β domains, which each have a 23 amino acid membrane-spanning domain. *In vivo* the binding of insulin initiates a sequence of events involving the autophorylation of tyrosine residues in the cytoplasmic domains of the β-subunits, the clustering of receptors, and the formation of clathrin-coated pits followed by internalization of cytoplasmic vesicles and the concomitant triggering of a signal that causes the plasma membrane to become enriched in sugar transporters (Carpentier, 1989). The lipid composition and fluidity of the bilayer will influence the rate of clustering and have some influence on the binding characteristics of the receptor *in vivo* (Ginsberg et al., 1981). It has also been found that the nature of the lipid headgroups influences the binding properties of the receptor by use of phospholipases to selectively hydrolyze phosphoglycerides and sphingolipids in whole cells (McCaleb and Donner, 1981). However, when the adipocyte insulin receptor is reconstituted in liposomes, the Gibbs energy of binding is found to be very similar to that for insulin binding to many different cell types, suggesting that there are no very large differences in binding characteristics with bilayer composition (Jones et al., 1986).

These observations are consistent with the peripheral position of the receptor binding domain and suggest that the choice of lipid for receptor reconstitution would not be expected to have a major effect on the receptor properties, provided the bilayer is in the liquid-crystalline state so that the receptors are uniformly distributed in the bilayer. In the gel state, the possibility of receptor clustering must be considered, as the receptors are "squeezed out" of the gel phase and laterally separated into receptor-rich clusters. Under these conditions, the binding properties may be markedly affected by lateral interactions between neighboring receptors.

8.4.3. The Effects of Cholesterol in Reconstitution

It is well established that cholesterol at sufficiently high levels (~ 33 mole %) in bilayers will eliminate the gel to liquid-crystalline phospholipid transition. By itself, cholesterol will not form bilayers but exerts its effect by interdigitation into phospholipid bilayers. At low levels (< 33 mole %), it reduces the fluidity of the bilayer below and increases the fluidity above the chain-melting temperature. It thus acts as a "plasticizer," so that the bilayer has an approximately constant fluidity over a wide temperature range. At low temperatures, cholesterol increases the apparent partial specific volume (V_a) of phospholipid, while at high temperatures it reduces it. The effect is illustrated in Figure 8.6, which shows V_a for DPPC-cholesterol bilayers as a function of both temperature and composition. The gel to liquid-crystalline transition at low levels of cholesterol is characterized by peaks in the V_a vs. cholesterol content (mole %) curves and rapid rises in V_a with temperature as the phospholipid goes through the chain-melting transition. Cholesterol has a fusogenic effect on unilamellar vesicles held a few degrees below the chain-melting temperature of their phospholipid, and when vesicles are incubated with membrane proteins at the temperature of optimum fusion they spontaneously incorporate protein. Figure 8.7

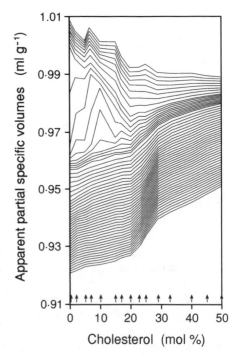

Fig. 8.6. Apparent partial specific volume (V_a) of dipalmitoylphosphatidylcholine-cholesterol bilayers as a function of cholesterol content (mole %) and temperature. The arrows denote the cholesterol concentrations used to construct the plot. At a constant cholesterol content, a vertical cut represents V_a vs. temperature. The lowest line corresponds to 0°C; the others are drawn for 1°C intervals up to 50°C. From Carruthers and Melchior (1986), with permission of the publisher.

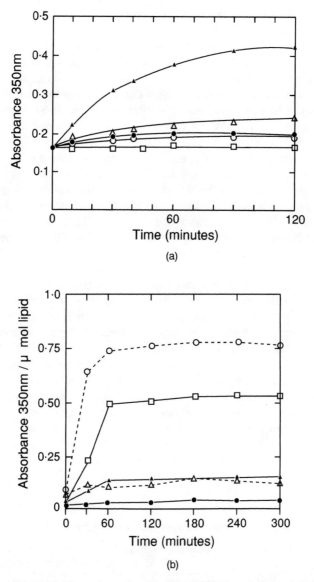

Fig. 8.7. a. Absorbance/time plots of unilamellar DMPC vesicles containing 12 mole % cholesterol after incubation at various temperatures. $\square$, 30°C; ▲, 21°C; △, 18°C; ●, 10°C; ○, 5°C. The absorbance measurements were made after incubation of the vesicles for 5 min. at 37°C to eliminate absorbance effects due to aggregation. **b.** Absorbance/time plots for unilamellar DMPC vesicles containing different concentrations of cholesterol after incubation at 21°C. Cholesterol mol % 3.9 (●), 7.5 (▲), 13.8 ($\square$), 19.9 (○), and 23.8 (△). The absorbance measurements were made after incubation of the vesicles for 10 min at 37°C, after which the samples were recooled to 21°C to continue the experiment. From Scotto and Zakim (1986), with permission of the American Chemical Society.

shows the effect of cholesterol on the fusion of dimyristoylphosphatidylcholine (DMPC) unilamellar vesicles as assessed from the absorbance changes of a vesicle suspension as a function of time (note that pure DMPC has a chain-melting temperature of 23°C). DMPC vesicles aggregate below the chain-melting temperature, giving rise to light scattering and an increased absorbance. In order to separate absorbance changes due to aggregate from those due to fusion, the absorbances were measured after incubation at 37°C for 5–10 min to dissociate any aggregates. In Figure 8.7a, it is seen that the absorbance increases with time of incubation in the temperature range 5–30°C but the increase is largest at 21°C. In Figure 8.7b the effect of fusion at 21°C is shown as a function of cholesterol content and demonstrates that there is an optimum cholesterol concentration in the region of 20% above which fusion decreases. The fusion process at 21°C facilitates the spontaneous incorporation of protein into the bilayer. Table 8.5 shows that on mixing DMPC vesicles with bacteriorhodopsin followed by incubation at 21°C for 1 or 10 min, the protein is spontaneously incorporated into the bilayer and that the extent of incorporation increases with cholesterol concentration. However, even at 0.1 mole % cholesterol, the amount of protein incorporation is significantly increased above that for pure DMPC vesicles; i.e., the cholesterol is facilitating (catalyzing) the incorporation. Other compounds, such as myristate and trace amounts of cholate, have also been shown to have an effect similar to cholesterol in catalyzing spontaneous protein incorporation (Scotto and Zakim, 1986).

The mechanism of the catalysis possibly relates to the fact that neither myristate nor cholesterol mixes ideally with phosphatidylcholines, so that in the gel state the bilayer will have localized packing defects that facilitate fusion and protein incorporation. This spontaneous incorporation of integral membrane proteins into preformed vesicle bilayers by the introduction of packing defects probably occurs during reconstitution by the freeze-thaw-sonication technique. It should be noted that when the surfactant is removed from many solubilized membrane proteins, they aggregate, possibly trapping residual sur-

TABLE 8.5. Cholesterol-Catalyzed Incorporation of Bacteriorhodopsin Into Unilamellar Dimyrstoylphosphatidylcholine Vesicles at 21°C

Concentration of Cholesterol (mol %)	Molar Ratio of Protein to Lipid	
	1 Min. Fusion Time	10 Min. Fusion Time
0	11	13
0.1	<20	32
1	<20	54
3	<20	104
6	56	97
12	99	134
18	102	222

From Scotto and Zakim (1986), with permission of the American Chemical Society.

factant, which like cholesterol catalyze the incorporation of the protein into the bilayer.

REFERENCES AND FURTHER READING

Beyhl, F.E. (1986) Interactions of detergents with microsomal enzymes: Effects of phospholipase C (*Bacillus cereus*) towards micellar short-chain phosphatidylcholine. IRCS Med. Sci. *14*, 417–418.

Brunner, J., Hauser, H. and Semenza, G. (1978) Single bilayer lipid-protein vesicles formed from phosphatidylcholine and small intestinal sucrose isomaltase. J. Biol. Chem. *253*, 7538–7546.

Carpentier, J.L. (1989) The cell biology of the insulin receptor. Diabetologia *32*, 627–635.

Carruthers, A. and Melchior, D.L. (1986) How bilayer lipids affect membrane protein activity. Trends Biochem. Sci. *11*, 331–335.

Coronado (1986) Recent advances in planar phospholipid bilayer techniques for monitoring ion channels. Annu. Rev. Biophys. Chem. *15*, 259–277.

El-Sayert, M.Y. and Roberts, M.F. (1985) Charged detergents enhance the activity of phospholipase C. (Bacillus cereus) towards micellar short-chain phosphatidylcholine. Biochim. Biophys. Acta *831*, 133–141.

Florine, K.I. and Feigenson, G.W. (1987) Protein redistribution in model membranes: Clearing of M13 coat protein from calcium-induced gel-phase regions in phosphatidylserine/phosphatidylcholine multilamellar vesicles. Biochemistry *26*, 2978–2983.

Garcia, M.L., Viitanen, P., Foster, D.L. and Kaback, H.R. (1983) Mechanism of lactose translocation in proteoliposomes reconstituted with lac carrier protein purified from *E. coli*. Part 1: Effect of pH and imposed membrane potential on efflux, exchange and counterflows. Biochemistry *22*, 2524–2531.

Ginsberg, B.H., Brown, T.J., Simon, I. and Spector, A.A. (1981). Effect of membrane lipid environment on the properties of insulin receptor. Diabetes *30*, 773–780.

Goodwin, G.C., Hammond, K., Lyle, I.G. and Jones, M.N. (1982) Lectin-mediated agglutination of liposomes containing glycophorin. Effects of acyl chain length. Biochim. Biophys. Acta *689*, 80–88.

Gorga, J.C. and Lienhard, G.E. (1984) One transporter per vesicle: Determination of the basis of insulin effect on glucose transport. Fed. Proc. (FASEB) *43*, 2237–2241.

Helenius, A., Sarvas, M. and Simons, K. (1981) Asymmetric and symmetric membrane reconstitution by detergent elimination. Studies with Semliki-Forest-virus spike glycoprotein and penicillinase from the membrane of *Bacillus licheniformis*. Eur. J. Biochem. *116*, 27–35.

Jackson, M.L. and Litman, B.J. (1982) Rhodopsin-phospholipid reconstitution by dialysis removal of octyl glucoside. Biochemistry *21*, 5601–5608.

Jain, M.K. and Zakin, D. (1987) The spontaneous incorporation of proteins in preformed bilayers. Biochim. Biophys. Acta *906*, 33–68.

Jones, M.N. and Nickson, J.K. (1981) Monosaccharide transport proteins of the human erythrocyte membrane. Biochim. Biophys. Acta *650*, 1–20.

Jones, M.N., Manley, P., Midgley, P.J.W. and Wilkinson, A.E. (1982) The dissociation of bovine and bacterial catalase by sodium *n*-dodecylsulphate. Biopolymers *21*, 1435–1451.

Jones, M.N., More, J.E. and Riley, D.J. (1986) A thermodynamic approach to hormone-receptor interaction: Application to insulin binding to adipocytes, adipocyte plasma membranes and liposomes incorporating adipocyte insulin receptors. J. Receptor Res. *6*, 361–380.

Jones, M.N., Finn, A., Mosavi-Movahedi and Waller, B.J. (1987a). The activation of *Aspergillus niger* catalase by sodium *n*-dodecylsulphate. Biochim. Biophys. Acta *913*, 395–398.

Jones, O.T., Earnest, J.P. and McNance, M.G. (1987b) Solubilization and reconstitution of membrane proteins in biological membranes: A practical approach. IRL Press, Oxford, chap. 5.

Kasahara, M. and Hinkle, P.C. (1977) Reconstitution and purification of the D-glucose transporter from human erythrocytes. J. Biol. Chem. *252*, 7384–7390.

Klausner, R.D., Bridges, K., Tsunoo, H., Blumenthal, R., Weinstein, J.N. and Ashwell, G. (1980) Reconstitution of the hepatic asialoglycoprotein receptor with phospholipid vesicles. Proc. Natl. Acad. Sci. USA *77*, 5087–5091.

Levitzki, A. (1985) Reconstitution of membrane receptor systems. Biochim. Biophys. Acta *822*, 127–153.

Lévy, D., Bluzat, A., Seigneuret, M. and Rigaud, J-L. (1990) A systematic study of liposome and proteoliposome reconstitution involving Bio-bead-mediated Triton X-100 removal. Biochim. Biophys. Acta *1025*, 179–190.

Maneri, L.R. and Low, P.S. (1988) Structural stability of the erythrocyte anion transporter, band 3, in different lipid environments. J. Biol. Chem. *263*, 16170–16178.

McCaleb, M.L. and Donner, D.B. (1981) Affinity of the hepatic insulin receptor is influenced by membrane phospholipids. J. Biol. Chem. *256*, 11051–11057.

McDonnel, A. and Staehelin, L.A. (1980) Adhesion between liposomes mediated by the chlorophyll a/b light scattering-harvesting complex isolated from chloroplast membranes. J. Cell. Biol. *84*, 40–56.

More, J.E. and Jones, M.N. (1983) The effect of membrane composition and alcohols on the insulin-sensitive reconstituted monosaccharide transport system of rat adipocyte plasma membranes. Biochem. J. *216*, 113–120.

Noel, H., Goswani, T. and Ponde, S.V. (1985) Solubilization and reconstitution of rat liver mitochondrial carnitine acylcarnitine translocase. Biochem. *24*, 4504–4509.

Pederson, S.E. and Ross, E.M. (1982) Functional reconstitution of β-adrenergic receptors and the stimulatory GTP-binding protein of adenylate cyclase. Proc. Natl. Acad. Sci. USA *79*, 7228–7232.

Pick, U. (1981) Liposomes with a large trapping capacity prepared by freezing and thawing of sonicated phospholipid mixtures. Arch. Biochem. Biophys. *212*, 186–194.

Radany, E.W., Bellet, R.A. and Garbes, D.L. (1985) The incorporation of a purified, membrane-bound form of guanylate cyclase into phospholipid vesicles and erythrocytes. Biochim. Biophys. Acta *812*, 695–701.

Reinhardt, R., Bridges, R.J., Rummel, W. and Lindemann, B. (1987) Properties of an anion-selective channel from rat colonic enterocyte plasma membranes reconstituted into planar phospholipid bilayers. J. Membr. Biol. *95*, 47–54.

Rivnay, B. and Metzger, H. (1982) Reconstitution of the receptor for immunoglobulin E into liposomes: Conditions for incorporation of the receptor into vesicles. J. Biol. Chem. *257*, 12800–12808.

Saito, S. and Tsuchiya, T. (1984) Characteristics of *n*-octyl β-D-thioglucopyranoside, a new non-ionic detergent useful for membrane biochemistry. Biochem. J. *222*, 829–832.

Scotto, A.W. and Zakim, D. (1986) Reconstitution of membrane proteins: Catalysis by cholesterol of insertion of integral membrane proteins into preformed lipid bilayers. Biochemistry *25*, 1555–1561.

Tefft, R.E., Carruthers, A. and Melchior, D.L. (1986) Reconstituted human sugar transporter activity is determined by bilayer lipid head groups. Biochem. *25*, 3709–3718.

Tiller, G.E., Mueller, T.J., Dockter, M.E. and Strive, W.G. (1984) Hydrogenation of Triton X-100 eliminates its fluorescence and ultraviolet light adsorption while preserving its detergent properties. Anal. Biochem. *141*, 262–266.

Wheeler, T.J. and Hauck, M.A. (1985) Reconstitution of the glucose transporter from bovine heart. Biochim. Biophys. Acta *818*, 171–182.

CHAPTER 9

PROTEIN TRANSLOCATION AND THE ANCHORING AND DISPOSITION OF PROTEINS

9.1. PROTEIN TRANSLOCATION ACROSS MEMBRANES

The cell is known to synthesize a large number of different polypeptides, some 10^3 to 10^4 in number. An important question is, what are the signals that indicate the position in the cell to which these proteins should go? Some proteins are inserted into a membrane while others need to pass through one or more membranes to reach their final destination. Certain membranes can translocate different proteins. These are termed translocation-competent membranes. Examples of these are the endoplasmic reticulum (ER), the peroxisomal membrane, the bacterial plasma membrane, the inner membrane of mitochondria, and the inner and thylakoid membranes of chloroplasts. The mitochondrial membrane can transport proteins in both directions, while the other membranes only in one direction. Membranes derived from the endoplasmic reticulum— e.g., the Golgi complex, secretory vesicles, endosomes, lysosomes, and the smooth ER membranes—are said to be translocation incompetent.

The way in which the proteins are directed to a target membrane is usually by means of a short stretch of amino acids at or near the NH_2 terminus of the protein, with the exception of many peroxisomal targeting signals that are at the COOH terminus. Most of these NH_2 terminal-targeting signals (called leader or signal sequences) are proteolytically removed by a signal peptidase on the *trans* side of the membrane. Apparently if the removal of the targeting signal is blocked, then translocation can still take place. Some translocated proteins carry targeting signals that are not proteolytically removed under normal conditions. To cross the ER and bacterial membrane, hydrophobic signals are used. These have a relatively hydrophobic NH_2 terminus with one or two basic residues, an apolar hydrophobic core of seven or eight residues, and a relatively hydrophilic COOH terminus ending with an amino acid carrying a small side chain. The hydrophobic signals are cleaved by proteases that are integral proteins (Verner and Shatz, 1988; Wickner and Lodish, 1985).

For targeting proteins into mitochondria and chloroplasts, hydrophilic sequences occur. These are rich in basic and hydroxylated residues and contain few if any acidic residues. They are reported to have no extended polar regions. The hydrophilic signals are removed by soluble proteases that require a metal $(Zn^{2+}, Mn^{2+}, or Co^{2+})$ as cofactors.

9.2. SIGNAL PEPTIDES

For the export of a protein—whether from yeast, higher eukaryotes, or bacteria—a general requirement is a signal sequence. Part of the recognition information in a nascent chain is in a contiguous sequence that specifies the eventual location of the mature protein, such as a signal sequence for export from the cell or a mitochondrial presequence for import into the mitochondria. These sequences can even be transplanted from one protein to another and still retain the same localization information (Gierasch, 1989).

These targeting sequences lack primary homology but have several properties in common. That is, they possess:

1. An amino terminal region with a net positive charge
2. A hydrophobic core of about 10 residues
3. Six to eight residues preceding the cleavage site, which often include proline or amino acids that favor turns
4. The cleavage site immediately follows the motif AXA in prokaryotes (where X is any amino acid). In eukaryotes, an amino acid with small side chains replaces Ala in this motif.

A schematic diagram is shown of a typical signal sequence in Figure 9.1.

The strong tendency for signal peptides to interact favorably with membranes, which argues for a direct membrane interaction *in vivo*, arises from the positive charge adjacent to a hydrophobic segment. These properties favor insertion into the endoplasmic reticulum or cytoplasmic membrane, subsequent to release of the nascent chain from the signal recognition particle (SRP) or

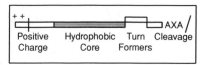

Fig. 9.1. Schematic diagram of signal peptide sequence.

SecA (Lee and Beckwith, 1986). This insertion would facilitate subsequent interactions with membrane proteins by restricting the nascent chain to two-dimensional diffusion and by presenting the hydrophobic portion of the signal sequence to binding sites on integral membrane proteins.

The amino-terminal charge and the hydrophobic core also appear to be involved in signal sequence binding to SRP. A model has been proposed based on the sequence of the 54 kDa subunit of SRP, which is known to bind the signal sequence in a methionine-rich pocket formed by helical segments of SRP (the so-called methionine-bristle binding site). Such a site could bind a variety of hydrophobic cores.

Many studies have been made of these signal sequences. There have been several spectroscopic studies using CD spectroscopy FTIR spectroscopy, NMR spectroscopy, and lipid monolayer systems (Briggs et al., 1986). Briggs and coworkers have examined the L and B wild-type signal sequence upon its interaction with a lipid membrane. These workers showed that the peptide adopted a β-type structure when associated with the surface of the lipid, but was predominantly α-helical when inserted into the lipid.

Some workers suggest that all export-competent signal peptides have the capacity to form an α-helix in interfacial and membrane mimetic environments. Furthermore, signal sequences spontaneously insert into the acyl chain region of phospholipids.

There is nevertheless still some uncertainty over the secondary structure of these signal peptides. Thus some workers find that the spectra of signal peptides show α-helical structures, while other workers find they adopt β-conformations. The signal sequence of the pho E gene product of *E. coli* indicates a preference for β-sheet conformation in micelles of Lubrol and lysolecithin. A β-conformation has also been observed for the signal sequence of *E. coli* λ-receptor in lipid monolayers (it is important to note that aggregation of peptides is quite common and could give rise to β-type structure).

In higher eukaryotes, where the components of the secretory apparatus have been more fully characterized (Rapoport, 1986), the first interaction of the signal sequence appears to be with the signal recognition particle. This interaction is probably the first committed step in protein secretion; it ensures, by virtue of the subsequent specific binding between SRP and its receptor in the ER membrane (SRP receptor or docking protein), that the nascent chain will be correctly targeted. Under some experimental conditions, SRP binding leads to an arrest or a pause in translation (Walter and Blobel, 1981), which is relieved by release of SRP upon its binding to SRP receptor (Gilmore and Blobel, 1983). A schematic diagram showing protein folding associated with membrane translocation is shown in Figure 9.2.

9.3. CHAPERONS

There is now known to be a class of proteins that can interact with a wide array of polypeptides from the time of synthesis to the final folded state. These

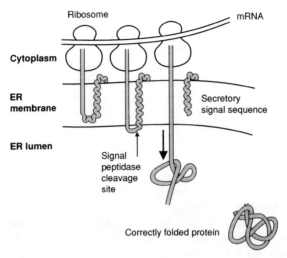

Fig. 9.2. Schematic diagram showing protein folding associated with translocation through the membrane. Most secretory proteins possess an N-terminal sequence of 16–30 residues (called the signal sequence). The secretory protein is synthesized on the ribosome and N-terminal amino acid(s) bind(s) the membrane of the endoplasmic reticulum (ER) using the signal recognition particle (SRP) and its receptor. A precursor protein cotranslationally moves through the ER membrane, and the signal sequence is cleaved by signal peptidase within the lumen of the ER. After translocation through the membrane, the secretory protein folds correctly according to the information encoded in its amino acid sequence.

proteins are known as chaperons. They had been previously studied and known as heat shock proteins (Hsps), which are synthesized by cells in response to stress such as an increase in temperature.

The Hsp 70 and the Hsp 60 families are linked to protein folding in cells. Most members of these families are present in cells under normal nonstressful conditions of the cell. The Hsp 70 family is derived from the fact that the first member is a 70 kD protein. The Hsp 60 proteins are organized as large assemblies arranged in two rings, each composed of seven 60 kD subunits. This link between the Hsps and normal protein synthesis has been demonstrated by studies of the interaction of cytoplasmic Hsp 70s with the nascent polypeptide chain while it is still on the ribosome. Studies with mitochondria have shown that the proteins directed to the endoplasmic reticulum or mitochondria are bound to Hsp 70 in the cytosol. Studies of an *in vitro* system consisting of the *E. coli* and Hsp 70, Hsp 60, and additional Hsps (GrpE, DnaJ, and GroES) have shown that the interaction of Hsps mimics the pathway for the interaction of Hsps in the *in vivo* import and folding of mitochondrial proteins. The Hsp 70 class of chaperons bind to unfolded and partially folded states of a variety of proteins but show little interaction with mature folded proteins. The release of bound proteins is influenced by an endogenous ATPase. In stressed cells, a plausible role for the Hsps induced by stress is to rescue unfolded or aggregated polypeptides back to an active conformation and the proteolysis of some denatured proteins.

Whether chaperons increase the rate of a limiting, on-pathway folding reaction or decrease the rate of off-pathway reactions, including those leading to aggregation, is still being considered (Agard, 1993).

9.4. PROTEIN INSERTION

Studies of various toxins indicate that they share a common requirement of insertion into or transport across the plasma membrane of the host cell. Some examples of this are tetanus toxin, diphtheria toxin, and *Staphylococcus* α-toxin that form pores in membranes.

Recent studies of colicins have led to a model for membrane insertion and channel formation. Colicins are plasmid-encoded proteins that kill bacteria and are in the range of 40–60 kDa molecular mass. The three-dimensional structure of the pore-forming domain of colicin has been deduced by X-ray crystallography. The protein consists of 10 α-helices organized in a three-layer structure. Two of the helices are completely buried within the structure and form a hydrophobic hairpin loop.

A schematic diagram according to Pattus et al. (1983) of the proposed mechanism of membrane insertion is shown in Figure 9.3. The first step is the initial interaction between the negatively charged lipids of the membrane and a ring

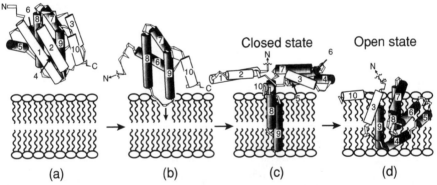

Fig. 9.3. A model for protein insertion into membranes suggested by the structure of the pore-forming domain of colicin A. Cylindrical rods represent α-helices. Light shading represents hydrophilic, and dark shading hydrophobic, portions of the protein surface. **a.** The initial interaction with the membrane. **b.** Spontaneous penetration of the hydrophobic hairpin. **c.** "Umbrella conformation.": the outer layers of the protein lie within the polar lipid headgroups, in contact with the hydrophobic core of the bilayer. For clarity they have been shown here above, rather than embedded into, the membrane surface. **d.** Application of a *trans*-negative membrane potential leads to the formation of the open channel. The channel may consist of a number of monomeric units. From Parker et al. (1990), with permission of the publisher.

of eight positively charged side chains positioned on a well-defined surface of the protein. The electrostatic field generated by the protein and membrane surfaces orientates the protein molecule so that the hydrophobic hairpin (helices 8 and 9) is positioned with its axis perpendicular to the membrane surface. The driving force for the initial insertion of the protein into the bilayer is considered to be the spontaneous penetration of the hairpin into the hydrophobic core of the membrane. This would leave the two charged outer layers of the protein embedded into the surface of the bilayer. The energy costs of disrupting the protein's three-dimensional fold is balanced by the replacement of hydrophobic interactions with the hydrophobic core of the bilayer. A low-pH environment optimizes the effect of the *trans*-negative potential required to form the open pore.

Relevance to protein transport. It is suggested that the common features of colicin membrane transport discussed above may provide a general transport mechanism for other proteins. Thus, for a protein to be transported across a membrane, it must be targeted to the correct membrane by specific interactions with protein receptors and then inserted or translocated across the membrane utilizing cellular energy.

9.5. LIPID-PROTEIN INTERACTIONS

The present consensus view of biomembrane structure is that the lipid bilayer is the basic matrix into which and around which the various proteins are situated.

The realization that integral proteins occur within the lipid bilayer structure led directly to the question of the extent to which the presence of such an integral protein perturbs the lipid environment and the significance that this may have on biomembrane structure and formation.

Following these early ideas, more sophisticated concepts developed concerning the perturbation of the layer of lipid adjacent to intrinsic proteins.

Electron spin resonance spectroscopy (ESR) experiments with cytochrome oxidase and sarcoplasmic reticulum ATPase (time scale $\sim 10^{-8}$ sec) showed the existence of "immobile components" in the corresponding spectra. It was proposed that the ESR immobile component is an indication of a special lipid shell (called the boundary).

Various terms were used to describe this perturbed lipid, including boundary-layer lipid, halo lipid, and annulus lipid. In all these views, it was assumed that a rigid or immobilized lipid shell exists separating the hydrophobic intrinsic protein from the adjacent fluid bilayer regions. It was suggested by some workers that this shell is a single rigid lipid layer. Others went further and argued that this single rigid lipid shell was long-lived as well as excluding cholesterol. It was suggested that the rate of exchange between this "annular lipid" and bulk lipid was slow even when the bulk lipid is in a fluid state.

Some workers suggested that the lipid perturbation extends to three layers or shells of lipid and others that it extended to the sixth or seventh lipid layer. Later workers Chapman et al. (1979) pointed to a number of observations that cast doubt upon this interpretation of the observed ESR "immobile" component and of the concept of special annulus lipid.

1. The observation of such an ESR "immobile" component is not restricted to proteins that may possess captive or bound lipids. It is indeed observed in gramicidin A/lipid/water systems. The gramicidin A molecule is a relatively simple polypeptide.

2. The immobile ESR component is observed at high protein (or polypeptide) content in the lipid bilayer. At such high protein or polypeptide content, the microviscosity will be high. Mobility of the probe molecule will be expected to be considerably inhibited, and multiple contacts of lipid with protein (trapped lipid) may occur.

3. NMR studies using either ^{1}H, ^{19}F, or ^{2}H nuclei (time scale 10^{-3}–10^{-5} s) on various biomembranes or reconstituted systems did not show the occurrence of two types of lipid. This is the case with ^{1}H NMR studies of rhodopsin in disc membranes; with ^{19}F NMR studies of E. coli membranes (Brown et al., 1977; Gent and Ho, 1978; Rice et al., 1979); and ^{2}H NMR studies of the cytochrome oxidase and sarcoplasmic reticulum systems.

Thus, as indicated by the NMR experiments, a continuity between the bulk lipid phase and the boundary layer lipids, and ready diffusion between these lipids, takes place.

Lipid specificity. No lipid specificity with respect to polar groups has been shown for several enzymes, such as (Na^+-K^+) ATPase, C_{55}-isoprenoid alcohol phosphokinase, or cytochrome c oxidase.

There are a few examples in which polar groups of phospholipids are connected to enzyme activity because they act like "allosteric effectors," producing a large increase in the enzyme activity by enhancing ligand binding (substrate or coenzyme). Such is the case of BDH and phosphatidylcholine; the cofactor NADH is not bound except in the presence of the lipid. Something similar is the case of pyruvate oxidase from E. coli. It has been also suggested that cardiolipin interaction with some components of mitochondrial ATPase may regulate its activity. Some other purified membrane enzymes have been examined for their preferences for polar groups like phosphoenolpyruvate phosphotransferase from bacterial membranes which would require phosphatidylglycerol for optimal activity; glucosyl and galactosyltransferases from *Salmonella typhimurium* require PE lipids. But Dean and Tanford (1977) showed that phospholipids can be fully substituted with exogenous, synthetic amphiphiles—e.g., detergents—without loss of enzyme activity with the sarcoplasmic reticulum ATPase.

It is important to consider in these situations the time scale appropriate to the particular physical technique used to study the protein-lipid perturbation. This can be appreciated when we realized that, measured on one time scale (say 10^{-8} s using ESR spin-labeled molecules), a molecule may appear to be rigid, while on another time scale (say 10^{-5} s using ^{2}H NMR methods), the same molecule can appear to be mobile. When attempts are made to relate some measured perturbation effects with enzymatic effects, then yet another time scale must be considered, i.e., the time interval over which the enzymatic conformational change occurs.

9.6. ANCHORING OF PERIPHERAL (EXTRINSIC) MEMBRANE PROTEINS

A large number of proteins have now been discovered that are anchored to plasma membranes via a glycosylphosphatidylinositol (GPI) grouping. This appears to be an alternative anchoring mechanism to those proteins that have a single pass hydrophobic transmembrane domain. GPI-anchored proteins have been found at most stages of eukaryotic evolution, including protozoa, yeast, slime, molds, invertebrates, and vertebrates (Fergusson and Williams, 1988). It appears to be the predominant form of anchoring cell surface proteins in protozoa. The basic structure of GPI membrane anchors is shown in Figure 9.4.

A range of sophisticated techniques have been used to examine the lipid components and the detailed structure of the carbohydrate structures that are present. Techniques such as delipidation, mass spectroscopy, high-resolution NMR spectroscopy (including two-dimensional NMR spectroscopy) have been applied in elegant studies to determine the structures of these anchoring molecules.

The lipid portions of GPI vary depending on their source. The alkyl/acyl chain component is reported to be unusual compared to normal cellular phosphatidylinositol phospholipids. The GPI anchors of *Trypanosoma brucci*, *T. congolense*, and *T. equiperdum* surface glycoproteins contain exclusively myristic (C 14:0) acid as *sn*-1,2 dimyristoylglycerol.

Studies of acetylcholinesterase (AChE) from erythrocytes show that G_2AChEs in mammalian erythrocytes are anchored to the extracellular cell plasma membrane. G_2AChEs with glycoinositolphospholipid linked covalently to the protein C terminus have been found in Torpedo electric organ insect heads and mammalian erythrocytes. These workers have shown an extra fatty acid (palmitic acid) is found in ester linkage to the inositol ring, and the presence of this extra fatty acid correlates with resistance to PI-PLC and *T. brucci* GPI-PLC enzymes. The biological functions that require this type of anchoring mechanism are still unclear. The role of glycosyl phosphoinositides in signal transduction processes has been discussed and a model involving the hydrolyses of glycosylphosphoinositides in plasma membranes with the interaction of insulin and its receptor in mind has been proposed (see Fig. 9.5).

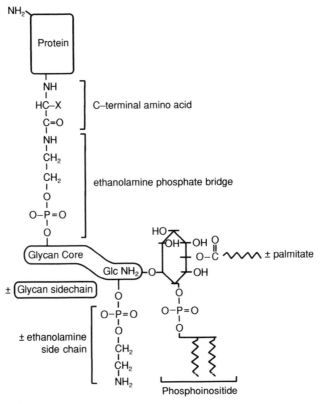

Fig. 9.4. The basic anatomy of GPI anchors. The $\pm$ symbol indicates groups that can be present or absent depending on the source of GPI. After Ferguson and Williams (1988).

The scrapie prion protein (Pr P^{Sc}) is also known to possess a glycosylphosphatidyl inositol (GPI) anchor. Scrapie is similar to the transmissible human encephalopathies Kuru, Creutzfeldt-Jakob disease, and Gerstmann-Straussler syndrome.

9.7. COVALENT ATTACHMENT OF FATTY ACIDS TO PROTEINS

There are now several examples of covalent attachment of fatty acids to proteins. In the first clear demonstration of this, the amino terminus of the Gram-negative lipoprotein constituent of the peptidoglycan cell wall was shown to be modified mainly with palmitic acid. Moreover, the side chain and the N-terminal residue cysteine were coupled to diacylglycerol via a thioether linkage. Membrane-bound penicillinases from Gram-positive bacteria appear to undergo the same kind of modifications.

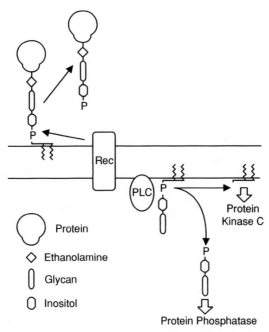

Fig. 9.5. The metabolism of glycosyl-PI in insulin action. A hypothetical model is presented illustrating the hydrolysis of glycosylphosphoinositides in plasma membranes. The interaction of insulin with its receptor causes the activation of receptor tyrosine kinase, probably the initial signal for receptor function. The activated receptor is then coupled by an unknown mechanism that may involve an intermediate G protein to the stimulation of one or more phospholipases C specific for glycosyl-PI. The hydrolysis results in the production of diacylglycerol that may cause a selective activation of the protein kinases C.

Hydroxylamine-stable incorporation of [^{14}C]myristic acid has been demonstrated in a number of viral and eukaryotic proteins (Towler et al., 1988). This stability to base/hydroxylamine treatment is taken to imply an amide link, almost certainly involving the amino-terminus, but the side chain of lysine is also a theoretical possibility. Furthermore, fatty acid attachment to the N-terminus is a highly specific process involving a glycine residue; mutagenesis of the N-terminal glycine of p60^{src} to alanine or glutamic acid residues abolished the modification. The attachment appears to occur during protein biosynthesis. Only one residue of myristic acid is attached per amino group. This myristic acid determines the membrane location of p60^{src}. It may be that membrane-associated proteins have additional sites that stabilize binding to the bilayer.

While most myristoylation is insensitive to hydroxylamine or base (e.g., 0.1 M KOH), attachment of most palmitic acid is, in contrast, sensitive to both these agents. Palmitoylation is not totally dependent on protein synthesis, implying that it is a late or turnover process. Examples of proteins that appear to undergo this modification include the ras oncogene family, viral membrane glycoproteins, ankyrin, the transferrin receptor, proteolipid protein from myelin, and rhodopsin. For p21^{ras}, palmitoylation is thought to be vital to membrane association, but for most other proteins the reasons for the formation of this modification are not easy to determine, since many of these proteins are integral by virtue of their primary structure and their activities appear not to be compromised by failure to carry out this modification. The attachment of either myristic or palmitic acid should not be regarded as inevitably indicating a membrane location (Magee et al., 1989).

Finally, it seems to be true that proteins which utilize fatty acid moieties as their only means of association with the membrane do not appear to directly involve the bilayer in any aspect of their activities other than simply localization.

REFERENCES AND FURTHER READING

Agard, D.A. (1993) To fold or not to fold. . . . Science *260*, 1903–1904.

Blobel, G. (1980) Intracellular protein topogenesis. Proc. Natl. Acad. Sci. USA *77*, 1496–1500.

Briggs, M.S., Cornell, D.G., Dloky, R.A. and Gierasch, L.M. (1986) Conformation of signal peptide induced by lipids suggests initial steps in protein export. Science *233*, 206–208.

Brown, M.F., Milijanich, G.P. and Dratz, E.A. (1977) Interpretation of 100- and 360-MHz proton magnetic resonance spectra of retinal rod outer segment disk membranes. Biochemistry *16*, 2640–2648.

Chapman, D., Gomez-Fernandez, J.C. and Goni, F.M. (1979) Intrinsic protein-lipid interactions: Physical and biochemical evidence. FEBS. Letts. *98*, 211–223.

Craig, E.A. (1993) Chaperones: Helpers along the pathways to protein folding. Science *260*, 1902–1903.

Dean, W.L. and Tanford, C. (1977) Reactivation of a lipid-depleted Ca^{2+}-ATPase by a nonionic detergent. J. Biol. Chem. *252*, 3551–3553.

Fergusson, M.A.J. and Williams, A. (1988) Cell surface anchoring of proteins via glycosyl-phosphatidyl inositol structures. Annu. Rev. Biochem. *57*, 285–320.

Gent, M.P.N. and Ho, C. (1978) Fluorine-19 nuclear magnetic resonance studies of lipid phase transitions in model and biological membranes. Biochemistry *17*, 3023–3038.

Gierasch, L.M. (1989) Signal sequences. Biochemistry *28*, 923–930.

Gilmore, R. and Blobel, G. (1983) Transient involvement of signal recognition particle and its receptor in the microsomal membrane prior to protein translocation. Cell *35*, 677–685.

Lee, C. and Beckwith, J. (1986) Cotranslational and posttranslational protein translocation into prokaryotic systems. Annu. Rev. Cell Biol. *2*, 315–336.

Magee, A.I., Gutierrez, L., Marshall, C.J. and Hancock, J.F. (1989) Targeting of oncoproteins to membranes by fatty acylation. J. Cell Suppl. *11*, 149–160.

Parker, M.W., Tucker, A.D., Tsernoglou, D. and Pattus, F. (1990) Insights into membrane insertion based on studies of colicin. Trends Biochem. Sci. *15*, 126–129.

Pattus, F., Martinez, M.C., Dargent, B., Cavard, D. and Verger, R. (1983) Interaction of Colicin A with phospholipid monolayers and liposomes. Biochemistry *22*, 5698–5703.

Rapoport, T.A. (1986) Protein translocation across and integration into membranes. CRC Crit. Rev. Biochem. *20*, 73–137.

Rice, D.M., Meadows, M.D., Scheinman, A.O., Goni, F.M., Gomez-Fernandez, J.C., Moscarello, M.A., Chapman, D. and Oldfield, E. (1979) Protein-lipid interactions: A nuclear magnetic resonance study of sarcoplasmic reticulum Ca^{2+}-ATPase, lipophilin and proteolipid apoprotein-lecithin systems and a comparison with the effects of cholesterol. Biochemistry *18*, 5893–5903.

Towler, D.A., Gordon, J.I., Adams, S.P. and Glaser, L. (1988) The biology and enzymology of eukaryotic protein acylation. Annu. Rev. Biochem. *57*, 69–99.

Turner, A.J., Ed. (1990) Molecular and Cell Biology of Membrane Proteins. Horwood, New York.

Verner, K. and Shatz, G. (1988) Protein translocations across membranes. Science *241*, 1307–1313.

Walter, P. and Blobel, G. (1981) Translocation of proteins across the endoplasmic reticulum. Part 3: Signal recognition protein (SRP) causes signal sequence-dependent and site-specific arrest of chain elongation that is released by microsomal membranes. J. Cell Biol. *91*, 557–561.

Wickner, W. (1979) The assembly of proteins into biological membranes: The membrane hypothesis. Annu. Rev. Biochem. *48*, 23–45.

Wickner, W. (1989) Secretion and membrane assembly. TIBS *14*, 280–283.

Wickner, W. and Lodish, H. (1985) Multiple mechanisms of protein insertion into and across membranes. Science *230*, 400–407.

CHAPTER 10

MEMBRANE DYNAMICS

10.1. MOLECULAR MOTION OF BIOMEMBRANES

10.1.1. Lipid Lateral Diffusion

The present view of biomembrane structure is a dynamic one in which the lipid components and protein components of many membranes are able to undergo considerable molecular motion. This includes the wagging and twisting of the CH_2 groups of the melted lipid chains above the chain melting temperature, T_c, and also the lateral and rotational diffusion of the lipids and the proteins within the plane of the lipid matrix.

Simple expressions have been derived to describe the diffusion of the lipid molecules. Thus the diffusion of a lipid molecule in the plane of the lipid bilayer is given by the equation (similar to Fick's second law of diffusion)

$$\frac{\delta c(x,t)}{\delta t} = D \frac{\delta^2 c(x,t)}{\delta x^2} \qquad (10.1)$$

where $c(x,t)$ is the concentration of diffusing molecules at the point x at time t. The diffusion coefficient D has dimensions $(length)^2/(time)$.

This can also be expressed in terms of the flux j where

$$j(x,t) = -D \frac{\delta c(x,t)}{\delta x} \qquad (10.2)$$

When the surface concentration is measured in moles per square centimeter, j is the number of molecules per second crossing a line of length 1 cm perpendicular to the direction x.

A number of measurements have been made of lipid diffusion coefficients using spin-labeled analogues within phospholipid bilayers. Träuble and Sackmann incorporated a spin-labeled steroid into dipalmitoyl phosphatidylcholine–water systems (Träuble and Sackmann, 1972), and for $T > T_c$ a diffusion coefficient of 1×10^{-8} cm^2/s was obtained. The diffusion coefficient of a spin-labeled phospholipid analogue in a sonicated egg lecithin–water system was measured and gave a value of $15 \pm 2 \times 10^{-8}$ cm^2/s at 40°C. Devaux and McConnell (1972) determined a lateral diffusion coefficient of $1.8 \pm 0.6 \times 10^{-8}$ cm^2/s. Other techniques have also been used, including NMR spectroscopic methods. Measurements of proton NMR spin-lattice relaxation times made possible the direct determination of phospholipid lateral diffusion coefficients. A diffusion coefficient of 3.8×10^{-9} cm^2/s was determined with a dipalmitoyl lecithin aqueous system, while studies based upon ^{31}P spectral line widths gave values of 1.8×10^{-8} cm^2/s at 30°C.

Monolayer systems have also been used for lipid lateral diffusion studies. Thus cholesterol diffusion as a function of surface pressure was measured using a radio isotope method. Depending upon the surface pressure, diffusion coefficients at 22°C have been determined that give values in the range 10^{-6} to 10^{-7} cm^2/s.

Some workers have used lipid probes and a fluorescent photo-bleaching technique for determining diffusion coefficients in lipid bilayer systems, while other workers have used triplet probes. Below the T_c temperature when the lipid is in the gel phase, a marked reduction in lipid lateral diffusion occurs.

10.1.2. Lipid Flip-flop

Yet another motion has been proposed, that lipid molecules can move from one side of the lipid bilayer to the other. This is sometimes termed flip-flop. In this case, the polar moiety is considered to pass through the hydrophobic region of the bilayer.

This process, where the lipid on one side of the lipid bilayer flips to the other side, is therefore expected to be relatively slow.

The first measurements of flip-flop were made by Kornberg and McConnell (1971) using a spin-labeled lecithin. The spin label was attached to the choline portion of the phospholipid molecule. The experiment consisted of preparing egg-yolk lecithin vesicles with a single lipid bilayer and containing an internal solvent-filled cavity. A small percentage of spin-labeled molecules was added to the lipid from which the vesicles were made, and was found to distribute itself more or less randomly between the outer and inner surface of the bilayer. The initial intensity of the electron spin resonance (ESR) spectroscopy decreased by 65% on addition of ascorbate, which reduced the spin label and had been shown previously not to enter the internal cavity of the vesicle. The rate of interchange between the two bilayer surfaces was then measured by following the appearance of more reducible label as a function of time; this was taken to represent labeled molecules that had crossed from the inner to the outer surface. The half-time of the process was found to be 6.5 hr at 30°C. Other, later experiments indicate that this lipid flip-flop occurs at a much slower rate—e.g., some workers used a specific exchange protein that binds phosphatidylcholine and catalyzes the rapid exchange between the bound lipid and phosphatidylcholine in lipid bilayer membranes. Using this approach, about two-thirds of the lipid in the phosphatidycholine vesicles was found to be accessible for immediate exchange. The appearance of additional exchangeable lipid (from the inside half of the bilayer), however, was not observed to take place at a measurable rate. This led to the conclusion that the half-time for lipid exchange must be greater than 11 days at 37°C. Other workers suggest the half-life time at 23°C for exchange is as long as 16 to 69 days.

10.2. MOLECULAR MOTION OF MEMBRANE PROTEINS

Membrane proteins contain portions that occur in the lipid matrix and other portions that appear in the aqueous environment. This means that a range of motions are possible. These include lateral diffusion, rotational diffusion, and also segmental motions of the amino acid groups.

10.2.1. Lateral Diffusion of Proteins

Membrane protein lateral diffusion has been studied using fluorescence photobleaching methods. With this technique it is necessary to be able to bleach a chromophore irreversibly with an intense light pulse. Measurements are made with a single cell using a microscope and the chromophore is usually fluorescein or rhodamine.

In this type of measurement, a small spot (some μm^2) on a single cell is irradiated by an intense laser pulse and the chromophores present in the spot are irreversibly bleached. The redistribution of the unbleached chromophores is then followed with the same laser beam attenuated some 10^4-fold. Depending on how the chromophores are characterized, the experimental procedure is called fluorescence or absorption microphotolysis. These are sometimes referred to as fluorescence recovery (or redistribution) after photobleaching (FRAP) and fluorescence photobleaching recovery (FPR). A variation on the method allows bleaching and redistribution to occur simultaneously. A small area of a cell under a microscope is irradiated, this time continuously with an intermediate intensity. The time course of the fluorescence signal originating from the irradiated area from its surroundings depends upon the rate at which chromophores are decomposed and unbleached chromophores enter the irradiated area. This procedure is sometimes called continuous fluorescence (or absorption) microphotolysis.

Rhodopsin was the first system for which quantitative lateral diffusion measurements were reported. This is because rhodopsin contains a suitable bleachable intrinsic chromophore. With an absorption microphotolysis method, Poo and Cone (1974) measured the lateral diffusion coefficient for rhodopsin within the frog and mudpuppy retina at 20°C, obtaining a value of $3.5 \pm 1.5 \times 10^{-9}$ $cm^2 s^{-1}$ and $3.9 \pm 1.2 \times 10^{-9}$ $cm^2 s^{-1}$, respectively.

Fluorescence microphotolysis measurements, applied to membranes other than the disc membrane, yield protein diffusion coefficients much smaller than those obtained for rhodopsin. Wey et al. (1981) labeled frog rod outer segments with 1A-tetramethylrhodamine, which reacted mainly with rhodopsin. The diffusion coefficient at 22°C was $(3.9 \pm 1.2) \, 10^{-9}$ $cm^2 s^{-1}$ if estimated from the rate of recovery of fluorescence in the bleached spot, and $(5.3 \pm 2.4) \, 10^{-9}$ $cm^2 s^{-1}$ if estimated from the rate of depletion of fluorescence from nearby regions.

The first lateral diffusion measurement with proteins that lack an intrinsic chromophore was performed with the human erythrocyte membrane (Peters et al., 1974). After labeling the hemoglobin-free erythrocyte ghosts with fluorescein-isothiocyanate, the protein lateral diffusion was studied by the fluorescence microphotolysis method. In contrast to the rhodopsin system, the lateral diffusion of integral proteins in these ghost erythrocyte membranes was found to be at 20°C, severely restricted ($D_1 = < 3 \times 10^{-12}$ $cm^2 s^{-1}$). In contrast to this value, however, when the lateral diffusion of mouse spherocytic erythrocyte membranes (i.e., erythrocyte variants deficient of spectrin) was measured, the lateral diffusion coefficient was found to be much higher ($D_1 = (2.5 \pm 0.6)$ 10^{-9} $cm^2 s^{-1}$). The spectrin skeleton appears to restrict the lateral movement of the integral proteins by direct linkage and/or trapping of proteins within the skeleton network.

The density of hormone receptors in the plasma membrane is very low, but it has been possible to perform fluorescence microphotolysis measurements on such systems by the synthesis of highly fluorescent hormone derivatives. Thus lactalbumin has been substituted with rhodamine at a molar ratio of 7 and coupled to the hormones insulin and epidermal growth factor. A nerve growth factor has also been labeled directly with rhodamine at a ratio of between 8 and 10. The receptors for these hormones were found to be diffusely distributed in the plasma membrane of cultured cells, giving D_1 values for the hormone-receptor complex at around 5×10^{-10} $cm^2 s^{-1}$, with mobile fractions in the range 0.4 to 0.8. The lateral diffusion coefficient of epidermal growth factor-receptor complexes on living human epidermoid carcinoma cells, labeled with tetramethylrhodamine-isothiocyanate, was found to be 3×10^{-10} cm^2/s.

The problem of the lateral and rotational diffusion of protein molecules in biomembranes has been studied (Saffman and Delbrück, 1975) by applying

TABLE 10.1. Lateral Diffusion Coefficients of Membrane Proteins*

Protein	Cell Type	D_l (m^2sec^{-1})
Rhodopsin	Frog retinal rod outer segment	3.5×10^{-13}
Acetylcholine receptor	Rat myotubules	10^{-16}
Epidermal growth factor receptor	Epidermoid carcinoma cells	7×10^{-14}
Epidermal growth factor receptor	Fibroblast	3×10^{-14}
Coat protein (in liposomes)	M-13 virus	7×10^{-13}
Concanavalin A receptor	Rat myotubes	3×10^{-15}
Agglutinin receptor	Wheat germ	$4-8 \times 10^{-15}$
Immunoglobulin E	Various cells	$2-7 \times 10^{-14}$
Band 3 protein	Erythrocyte	4×10^{-15}
Bacteriorhodopsin (in liposomes)		10^{-12}
Bacteriorhodpsin (in purple membranes)		10^{-14}
ADP/ATP carrier	Mitochondria	2×10^{-13}
Cytochrome c	Mitochondria	10^{-12}

*Quinn and Chapman (1991).

classical analysis of Brownian motion to a hydrodynamic model. Saffman and Delbrück treated the protein molecules as a cylinder with an axis perpendicular to the plane of the membrane. They deduced an expression for the lateral diffusion coefficient.

$$D_l = \frac{k_BT}{4\pi\mu h}\left[\log\left(\frac{\mu h}{\mu_w r}\gamma\right)\right] \tag{10.3}$$

where μ is the membrane viscosity, μ_w is the aqueous phase viscosity, h is the height of the (cylindrical) protein molecule, r is its radius, γ is the Eulers constant, and k is the Boltzmann constant.

Putting reasonable values in this expression—say, $\mu = 0.2$ Pa sec and $\mu_w = 0.01$ P, h = 5 nm, r = 2 nm and $\gamma = 0.5722$—the diffusion coefficient would be $\sim 7 \times 10^{-13}$ m^2sec^{-1}. This figure is in reasonable agreement with the experimental values for protein lateral diffusion shown in Table 10.1. For the rotational mobility, they deduced an expression:

$$D_r = \frac{k_BT}{4\pi\mu r^2 h} \tag{10.4}$$

From equations 10.3 and 10.4, it can be seen that the dependence of lateral mobility on protein diameter is small compared with its effect on rotational mobility, where D_r is proportional to r^2.

10.2.2. Rotational Diffusion of Membrane Proteins

The first measurement of the rotational diffusion of a membrane protein was carried out with rhodopsin using the internal retinal chromophore and using the technique of laser flash photolysis (Cone, 1972). This measurement gave a value for the rotational correlation time of 20 μsec. This measurement was followed by studies on bacteriorhodopsin using the retinal internal membrane chromophore (Razi-Naqvi et al., 1973). This showed that by contrast with rhodopsin, bacteriorhodopsin was relatively immobile within the lipid matrix. The technique used a polarized pulse of light, and the decay of the induced dichroism was determined.

Following these studies, the extension of measurements to other membrane proteins was made possible by the introduction of triplet probes using molecules such as eosin. A triplet state is necessary because of the relatively slow rotational motion of the proteins within the lipid matrix (Razi-Naqvi et al., 1973).

While the rotational diffusion of macromolecules in aqueous solution has been studied by fluorescence depolarization methods with diffusion times in the nanosecond time scale, the viscous environment of the lipid matrix causes membrane protein rotational correlation times to be in the microsecond time scale.

The basic components of a flash photolysis apparatus for measuring the rotational diffusion of proteins by absorption depolarization is as follows. The exciting source is a laser with a light beam output energy of 10 to 150 mJ per flash, which is vertically polarized. This may be either a dye laser or a frequency-doubled neodymium laser. The advantage of the former is that it allows easy alteration of the excitation wavelength (fixed excitation wavelength of 530 nm for the neodymium laser), while the advantage of the latter is its short pulse width of about 20 nsec (compared with a few μsec pulse width for the dye laser).

If triplet probes are used, the sample in a fluorimeter cuvette should be gassed with pure nitrogen or argon in order to remove oxygen, which can quench the triplet states. Furthermore, removal of oxygen prevents photooxidation of, for example, $(Ca^{2+} + Mg^{2+})$ ATPase.

Absorption changes in the sample initiated by laser flash excitation are followed with a continuous measuring white light beam (50–100 W light intensity). After passing through the sample, the light is focused at the entrance slit of a monochomator. The light emerging from the exit slit is divided into parallel and perpendicular components with respect to the polarization of the exciting flash with the aid of a polarizing beam splitter. Each of the two components passes through a polarizing filter orientated with its polarizing axis according to its relation to the beam splitter. Light detection for the two components is

TABLE 10.2. Correlation Times of Rotational Diffusion of Membrane Proteins*

Protein	Source	t_c (μ sec)
Rhodopsin	Retinal rod outer segment	20
Cytochrome a_3	Mitochondrial inner membrane	500
Cytochrome P-450	Microsomal membrane	270
Cytochrome b_5	Microsomal membrane	0.4
Band 3 protein	Erythrocyte	4,000
Ca^{2+}-ATPase	Sarcoplasmic reticulum	70, 130, 200 (different methods below T_c)
Acetylcholine receptor	*Torpedo* electroplax	0.7
Glycophorin	Erythrocyte	1–2 (monomer) 10–20 (dimer)
Epidermal growth factor receptor	Human epidermoid carcinoma cell	350
Bacteriorhodopsin	*Halobacterium* purple membrane	20,000

*Quinn and Chapman (1991).

measured simultaneously with two photomultipliers. The photomultiplier gains are most easily matched by adjustment of their separate high-voltage supplies. The two signals from photomultiplier tubes can then be recorded and stored in a transient recorder with suitable time resolution. Some values of the rotational correlation times for a number of membrane proteins are given in Table 10.2. Membrane protein rotational correlation times may also be measured by using saturated transfer electron spin resonance techniques.

10.2.3. Flippase Proteins

P-glycoproteins pump drugs out of cells by an ATP-dependent process, thereby reducing their toxicity. Recently it has been suggested that the unusual properties of P-glycoprotein can be explained by a "flippase" model. The idea is that a drug molecule, rather than interacting with the transporter directly from the aqueous phase, interacts instead with the lipid bilayer. The protein is then envisaged to flip the drug from the inner leaflet to the outer layer (Higgins and Gottesman, 1992).

P-glycoprotein substrates are primarily cationic lipid-soluble planar molecules that can intercalate among the phospholipid molecules. Other "flippase" proteins are thought to be involved in translocating specific lipids from one leaflet of a lipid bilayer to the other in order to maintain lipid asymmetry within the biomembrane (see Chap. 7). The proteins responsible for "flippase" activity have not as yet been fully characterized but appear to require ATP hydrolysis (Devaux, 1988).

10.3. TRANSMEMBRANE SIGNALING AND RECEPTOR PROCESSING

10.3.1. Transmembrane Signaling

In recent years, considerable progress has been made towards an understanding of how cells sense and respond to external stimuli, i.e., how the signals are generated and transduced across the cell membrane. A fundamental principle of transmembrane signaling is that the stimulation tends to be highly specific even though there is a diversity of signals. Receptors for hormones, drugs, growth factors, light, and smell have all been characterized.

The ways in which ligand binding produces a cellular response are varied. The mechanisms include the modulation of adenylate cyclase activity, the facilitated diffusion of solutes and ions through gated channels, and also the activation of phospholipase activity to enhance the turnover of specific membrane phospholipids.

Adenylate cyclase is an integral protein of the plasma membrane; its activity is allosterically modulated by ligands bound to receptors on the outer surface of the membrane and the catalytic site is located on the cytoplasmic surface. The cytoplasmic concentration of the cyclic nucleotide (cyclic $3^1,5^1$-AMP) is regulated by synthesis, catalyzed by adenylate cyclase and by its metabolism to AMP mediated by the cytoplasmic enzyme phosphodiesterase. There are two types of receptors modulating adenylate cyclase activity—these are stimulatory and inhibitory.

Proteins designated as G proteins because they bind GTP are involved in cell signaling. One group of G proteins mediate a stimulation of adenylate cyclase activity and are coupled to activation of β-adrenergic, scrotonin, and glucagon receptors (see Fig. 10.1). Another group of G proteins is associated with muscarinic cholinergic receptor activation. Transducin is a special type of G protein that couples the light receptor in the visual process with generation of the nerve impulse.

10.3.2. Endocytosis and Receptor Processing

Receptors of all types are continually removed and replenished on the cell surface. The ligand-receptor complexes are segregated within lateral domains on the surface of the plasma membrane and concentrated into clathrin-coated pits. Subsequent internalization of these plasma membrane domains as coated vesicles results in a series of processing steps that depend on the nature of the ligand-receptor complex. On transfer to a prelysosomal compartment of low pH, some ligand-receptor complexes apparently dissociate, and many are known to recycle back to the cell surface where they are available to undergo further cycles of binding and internalization, although this is not always the case.

Transferrin is an example of a receptor of this type. This is an 80-kDa serine glycoprotein responsible for the uptake of iron into cells. The transferrin acts

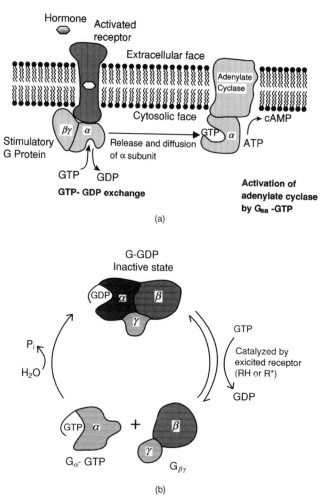

Fig. 10.1. a. The activation of adenylate cyclase by the binding of a hormone to its specific receptor is mediated by G_s, the stimulatory G protein. A single hormone-receptor complex catalyzes the formation of many molecules of G_s. **b.** The hydrolysis of GTP bound to the α-subunit of G_s terminates the activation of adenylate cyclase.

as the vehicle by which iron is internalized by clathrin-coated vesicle formation in a cyclic process whereby iron-free transferrin is subsequently secreted from the cell where it is again available to transport iron.

Hormones and growth factors are sometimes characterized by degradation of the ligand-induced internalized receptor complex.

The components of the plasma membrane are sorted prior to internalization with receptor-mediated endocytosis. The receptor-mediated endocytosis process is illustrated in Figure 10.2. It can be seen that the membrane becomes

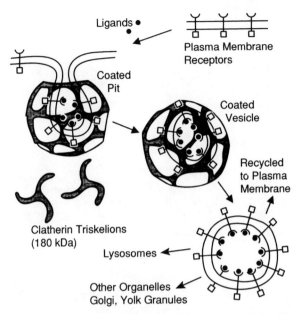

Fig. 10.2. Receptor-mediated endocytosis in clathrin-coated vesicles (Quinn and Chapman (1991)).

invaginated to form a pit, which is eventually internalized as a clathrin-coated vesicle that fuses with endosomal elements in the peripheral cytoplasm. These include vacuolar structures connected to tubular cysternae, and in the vacuoles small vesicles 30 nm in diameter are often observed.

The cells of the immune system are able to recognize a large variety of foreign macromolecules. The process is mediated through the T and B lymphocytes, which are characterized by expression of diverse sets of antigen receptors on the respective cell surfaces. The receptors on the B cells are antibodies. The T cells require the antigen to be processed and presented in an appropriate form before a response is elicited.

The molecular architecture of receptors involved in immune reactions is related and is immunoglobulin in type. A diagram of the various members of the immunoglobulin superfamily of receptors is shown in Figure 10.3. This shows receptors for polymeric IgA, neural cell adhesion molecule (N-CAM) receptor, T-cell antigen receptor, and TH_y-1 receptor. The polymeric IgA receptors are involved in the uptake of IgA from the blood by various glandular cells.

Cell adhesion molecules are also receptorlike and are Ca^{2+} dependent. Subclasses have been characterized in liver (L-CAM) and neural (N-CAM) cells. Adhesion between cells is homophilic, i.e., the N-CAM on one cell surface interacts with the identical N-CAM on another cell.

Another group of receptors of the plasma membrane create transmembrane

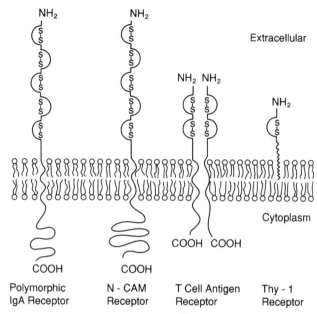

| Polymorphic
IgA Receptor | N - CAM
Receptor | T Cell Antigen
Receptor | Thy - 1
Receptor |

Fig. 10.3. Schematic representation of the immunoglobulin superfamily of cell surface receptors (Quinn and Chapman, 1991).

links between the extracellular matrix and components of the cytoplasm, e.g., actin, vinculin, thallin. Integrins are also concerned with cell–cell adhesion and cell–matrix interactions.

REFERENCES AND FURTHER READING

Cone, R.A. (1972) Rotational diffusion of rhodopsin in the visual receptor membrane. Nature *236*, 39–43.

Devaux, P. (1988) Phospholipid flippases. FEBs Lett. *234*, 8–12.

Devaux, P. and McConnell, H.M. (1972) Lateral diffusion in spin-labeled phosphatidylcholine multilayers. J. Am. Chem. Soc. *94*, 4475–4481.

Higgins, C.F. and Gottesman, M.M. (1992) Is the multidrug transporter a flippase? [Review] TIBS *17*, 18–21.

Kornberg, R.D. and McConnell, H.M. (1971) Lateral diffusion of phospholipids in a vesicle membrane. Proc. Natl. Acad. Sci. USA *68*, 2564–2568.

Peters, R., Peters, J., Tews, K.H. and Böhr, W. (1974) A microfluorometric study of translational diffusion in erythrocyte membranes. Biochim. Biophys. Acta *367*, 282–294.

Poo, M. and Cone, R.A. (1974) Lateral diffusion of rhodopsin in the photoreceptor membrane. Nature *247*, 438–441.

Quinn, P.J. and Chapman, D. (1991) The plasma membrane, in Structural Biology. E.E. Bitlar, Ed. JAI Press, pp. 1–75.

Razi-Naqvi, K., Gonzalez-Rodriguez, J., Cherry, R.J. and Chapman, D. (1973) Spectroscopic technique for studying protein rotation in membranes. Nature *245*, 249–251.

Saffman, P.G. and Delbrück, M. (1975) Brownian motion in biological membranes. Proc. Natl. Acad. Sci. USA *72*, 3111–3113.

Träuble, H. and Sackmann, E. (1972) Studies of the crystalline-liquid phase transition of lipid model membranes. Part 3: Structure of a steroid-lecithin system below and above the lipid phase transition. J. Am. Chem. Soc. *94*, 4499–4510.

Wey, C.L., Cone, R.H. and Edidin, M.A. (1981) Lateral diffusion of rhodopsin in photoreceptor cells measured by fluorescence photobleaching and recovery. Biophys. J. *33*, 225–232.

INDEX

Acetylcholine receptor, 238
Acetylcholine reconstitution study, 206
Acholeplasma sp., 184, 192
Acquired immunodeficiency syndrome
    (AIDS), 122–123
Acidity. *See* pH
Actin filaments, 193–194
Acyl carnitine reconstitution study, 207
Acyl chains
    length
        and cmc, 75
        and glucose transport, 210–211
        and phospholipid monolayers, 31–35
    of phospholipids, 10, 11
    unsaturation and chain melting, 105
Adenylate cyclase, 239
Adhesion molecules, 241–242
Adipocyte D-glucose transporter, 205
Adipocytes, and insulin transport, 213–214
β-Adrenergic receptor reconstitution study,
    206
Adsorption
    diffusion-controlled, 52
    with energy barrier, 52–54
    kinetics of, 50–51
    liposomal to cell surface, 134–135
    in protein monolayers, 50–54
AIDS (acquired immunodeficiency
    syndrome), 122–123
Alkalinity. *See* pH

Alkyl chains, membrane–protein interactions,
    17
Alpha-helix random coil model, 168
AmBisome, 122
Amide I band, 196
Amino acids; *see also* Proteins
    Gibbs energies, 14–17
    polarity/nonpolarity, 13
    polarity scale of residues, 18–19
Amphipathicity
    four-point charge model, 2–3
    principles of, 1–4
Amphipathic molecules
    in biomembranes, 17–20
    proteins as, 13–17
    in protein targeting, 20
    types, 5–17
        membrane lipids, 1–13
        natural surfactants, 8–10
        synthetic surfactants, 5–8
Amphipathic molecules, micellization of
    mixed systems, 95–98
Amphotericin B, 122; *see also* Liposomal
    targeting
Anchoring of proteins, 20, 64, 200, 227–228
Anionic surfactants, 6
Ankyrin, 193–194
Annulus (perturbed) lipids, 225–227
Antibody analysis of membrane proteins, 188
Antibody binding to liposomes, 133

Anti-infective drugs, 124–125
Apocytochrome c, 171–172
Archaebacterial lipids, 181–182
Arthritis, 126
Asialoprotein reconstitution study, 206
Asthma, 126
Asymmetrical distribution of lipids, 186–187
ATPases, 205, 226

Bacteriorhodopsin, 196–197, 205, 238
Beta-adrenergic receptor reconstitution study,
    206
Bilayer concept of biomembrane, 185
Bilayer planar lipid membranes, 202
Bilayers
    as back-to-back monolayers, 25
    equivalence with monolayers, 57–60
    hydrophobic–lyophilic balance, 144–146
    monolayer transition to surface, 35–36
    reconstitution studies, 199–217; see also
        Reconstitution
    surfactant interactions, 144–148
Bile acids, critical micelle concentrations
    (cmc), 71–74
Bile salts, 8–10
    micellar aggregation numbers, 84
    micellar size and shape, 81–84
    physiological function, 97
Binding
    equilibrium constants, 156–157
    Scatchard analysis, 157–161
    statistical effects, 164–166
    surfactant and protein denaturation, 151–
        153
    Wyman's binding potential (π) concept,
        162–164
Biomembranes
    amphipathic molecules in, 17–20
    bilayer concept, 185
    cytoskeleton, 192–193
    dynamics of, 232–243
    fluidity of bilayer matrix, 105
    fluid-mosaic model, 185–186
    human erythrocyte. See Human
        erythrocyte membrane
    integral and peripheral proteins, 192–194
    ion channels and receptor proteins, 189–
        190
    lipid composition, 180–183, 186–187
    and lipid phase transitions, 190–192
    molecular motion
        lipids
            flip-flop, 233–234
            lateral diffusion, 232–233

    proteins
        lateral diffusion, 234–237
        rotational diffusion, 237–238
    planar bilayer lipid, 202
    protein composition, 183–184, 187–188
    reconstitution studies, 199–217; see also
        Reconstitution
    spectroscopic studies, 195–197
    static and dynamic lipid asymmetry, 186–
        187
    surfactant-induced solubilization, 147–149
    surfactant interactions
        with lipids, 144–146
        in membrane solubilization, 147–149
        principles, 143–144
    transmembrane signaling, 189–190, 212–
        214, 239–242
Blood plasma proteins, and liposomal drug
    delivery, 138–139
Bonding, hydrophobic, 4
Boundary-layer (perturbed) lipids, 225–227
Bovine heart glucose transporter, 205

Ca^{2+}
    casein binding of, 78
    and phospholipid monolayer, 36–39
    and surface pressure, 111
Ca^{2+}-ATPase, 238
Calorimetry, 154
Carboxyfluorescein, as marker, 145–146
Caseins
    micellar-like behavior in, 76–79
    micellar size and shape, 86–88
    surface pressure, 55
Cationic surfactants, 6
Cell adhesion molecules, 241–242
Chain length
    and cmc, 75
    and glucose transport, 210–211
    in phospholipid monolayers, 31–35
Chain melting, 32, 102, 105–112
Chaperon proteins, 222–224
CHAPS surfactant, 9–10
Chlorophyll a/b light harvesting complex,
    207
Cholesterols
    crystallization, 45, 46
    in human erythrocyte membrane, 44
    and lipid chain crystallization, 107–110
    and lipid fluidity, 110
    and reconstitution, 214–217
    as surfactants, 8–10
    temperature and monolayer formation,
        44–45

Cholic acid, 8–9
Circular dichroism (CD) spectroscopy, 195, 222
Cluster model of Frank and Wen, 2
CMW method, 33–35
[^{14}C]Myristic acid, 229–230
Colicins, 224–225
Complementarity, hydrophobic, 212
Concanavalin A–bearing liposomes, 132–133
Conductivity
    and micellization, 66–67
    Onsager theory, 65–66
Contrast matched water condition (CMW method), 33–35
Cooperativity, of endothermic transition, 103–105
Counterion binding, 91
c regions, 20–21
Critical micelle concentration (cmc)
    of bile acids, 71–74
    of lecithin, 74–75
    of natural surfactants, 71–74
    of phospholipids, 74–75
    principles and concept, 64–67
    of synthetic surfactants, 67–71
Cryobiologic techniques, 191
Crystallization, 105–107
    cholesterol and, 107–110
Crystals
    order profile parameter, 107
    renatured vs. native cross-linked, 167–168
Cubic phases, 113–115
Cytochrome oxidase reconstitution study, 207
Cytochromes, 238
Cytoskeleton, 192–193

Danielli–Davson bilayer model, 185
Decorated micelle model, 170–172
Denaturation, 150–153
Diabetes mellitus. See Glucose transport; Insulin
Dialysis
    quantitative equilibrium, 153–155
    for reconstitution studies, 203
Diffusion
    lipid
        flip-flop, 233–234
        lateral, 232–233
    protein
        flippases, 238
        lateral, 234–237
        rotational, 237–238
Dimyristoylphosphatidic acid. See DMPA
Dimyristoylphosphatidylcholine. See DMPC

Dipalmitoylphosphatidylcholine. See DPPC
DMPA, monolayer ion/electrostatic interactions, 36–39
DMPC
    air–water interface studies, 33–35
    and glucose transport, 210
    and reconstitution, 214–217
n-Dodecyltrimethylammonium bromide
    micellization parameters, 94
    temperature and micellization, 95
Doxorubicin, 122; see also Liposomal targeting
DPPC
    and reconstitution, 214–217
    surface pressure, 57
Drug targeting, 124–139; see also Liposomal targeting

Electric field effects, 111–112
Electron spin resonance (ESR) spectroscopy, 225, 227
Electrophoresis. See SDS-PAGE technique
Electrostatic interactions, in DMPA monolayer, 36–39
Endocytosis, 135–139, 239–240
Endoplasmic reticulum (ER), 220–221
Enthalpy; see also Thermodynamics
    of chain transitions, 105–107
    of micellization, 93–95
Entropy, 14; see also Thermodynamics
    of chain transitions, 105–107
    of micellization, 93–95
Environmental effects of surfactants, 98
Enzyme kinetic analysis, 154
Epidermal growth factor receptor, 238
Equilibrium constants, of ligand binding, 156–157
Erythrocyte anion transporter, 205
Erythrocyte D-glucose transporter, 205
Erythrocytes. See Human erythrocyte membrane
Ethanol, and surfactant cmc, 71
Ethanol injection method, 119–120
Ether injection method, 120
Ether lipids, 180, 192
Eubacterial lipids, 180
Extrinsic proteins, 187, 188, 226–227

Fatty acids; see also Lipids; Phospholipids
    covalent attachment to proteins, 228–230
FLAP technique, 234–235
Flash photolysis, 237
Flexible helix model, 168
Flickering clusters model, 2

Flip-flop of lipids, 233–234
Flippase proteins, 238
Fluid-mosaic model of biomembrane, 185–186
Fluorescence recovery after photobleaching (FLAP technique), 234–235
Fluorescent probe studies
  bile salts, 82–83
  caseins, 86–88
  cholesterol, 48
  defined, 155
  disadvantages, 30–31
  of monolayers, 28–29, 29–31
Fourier transform infrared (FTIR) spectroscopy, 195–196, 222
Frank and Wen flickering clusters model, 2
Freeze-thaw-sonication, 202
FTIR spectroscopy, 195–196, 222

Gel structures, packing behavior, 105
Gibbs energy, 5; see also Thermodynamics
  of amino acids, 14–17
  of insulin binding, 213
  of micellization, 89–93
  of protein binding, 164–166
Gibbs-Helmholtz equation, 93–94
Gibbs monolayers, 24–25
Glucose transport, 205, 208–212
Glycerol, structure, 12
Glycophorins, reconstitution studies, 212
Glycophosphoinositol groupings (GTIs), 227–228
Glycoproteins
  O-linked vs. N-linked, 184
  reconstitution study, 204–214, 204–217
  SDS-PAGE analysis, 150
Glycophorin reconstitution study, 207
Gorter–Grendel lipid bilayer hypothesis, 185–186
G proteins, 239
Guanylate cyclase reconstitution study, 207

Halo (perturbed) lipids, 225–227
Halobacterium halobium, 196–197
Heat shock (chaperon) proteins, 222–224
Helix models
  flexible, 168
  α-helix random coil, 168
Hexagonal phases, 112, 191–192
Holobacterium halobrium, 196–197
h regions, 20–21
Hsp. See Heat shock (chaperon) proteins
Human erythrocyte membrane, 44, 186, 192–194

glucose transport, 209
glycophorin reconstitution study, 207
and hydrophobic complementarity, 212
lateral diffusion studies, 235
Hydrophilicity, 2
Hydrophobic bonding, 4, 20
Hydrophobic complementarity, 212
Hydrophobicity, 2
Hydrophobic–lyophilic balance, 144–146

Immune system
  and liposomal drug delivery, 135–139
  receptor processing in, 241
Immunoglobulin E reconstitution study, 206
Immunoglobulins, 206, 241–242
Immunoliposomes, 127, 132–133; see also Liposomal targeting
Insulin transport, 206, 213–214
Integrins, 242
Intrinsic proteins, 187, 188
Ion channels
  and receptor proteins, 189–190
  voltage-gated, 189–190

Lactose, 205
Lamellar phase, 103–110, 190
Langmuir-Adams surface balance, 25–27
Langmuir trough studies, 25–28
Laser flash photolysis, 237
Lecithin
  asymmetrical lipid distribution, 186–187
  critical micelle concentrations, 74–75
  flip-flop phenomenon, 233–234
  lyotropic mesomorphism and, 102–115
  micellization parameters, 94
Ligand binding. See Binding
Ligand-mediated liposomal targeting, 126–134
Light scatter techniques. See Fluorescent probe studies
Lipid bilayer. See Bilayers
Lipid hydrated states and phase behavior, 102–115, 190–192; see also under Phospholipids
Lipid perturbation, 225–227
Lipids; see also Phospholipids
  annulus (perturbed), 225–227
  archaebacterial, 181–182
  asymmetrical distribution, 186–187
  biomembrane composition, 180–183, 186–187
  boundary-layer (perturbed), 225–227
  eubacterial, 180–182
  flip-flop, 233–234

halo (perturbed), 225–227
lateral diffusion, 232–233
perturbed, 225–227
receptor function, 212–214
specificity of, 226–227
static and dynamic asymmetry of
        biomembrane, 186–187
Lipid specificity, 226–227
Liposomal targeting; *see also* Liposomes
and bilayer–surfactant interactions, 144–
        148
compartmental, 126
ligand-mediated (active), 126–134
and liposome–cell interactions, 134–136
natural (passive), 124
physical, 124–126
and reticuloendothelial system, 136–139
Liposomes; *see also* Liposomal targeting
as carriers, 121–124
cellular interactions, 134–136
formation
        preparation methods, 118–120
        stability, 120–121
        types, 117–118
pH-sensitive, 124–125
and reconstitution, 214–217
temperature-sensitive, 124–125
types
        intermediate-sized unilamellar vesicles
                (IUVs), 118
        large unilamellar vesicles (LUVs), 118
        multilamellar vesicles (MLVs), 118
        small unilamellar vesicles (SUVs),
                117–118
Liquids, physical properties of common, 2
Lyotropic mesomorphism, 102–115
cubic phases, 113–115
hexagonal phases, 112, 191–192
lamellar phase, 103–110, 190
phase separation, 110–112

"Magic bullet" hypothesis, 121
M13 bacteriophage coat protein reconstitution
        study, 207
Media, composition and cmc of synthetic
        surfactants, 67–71
Membrane. *See* Biomembrane
Membrane lipids, 10–13; *see also*
        Phospholipids
Mesomorphism
lyotropic, 102–115
thermotropic, 102–107
Micellar-complex model, 168

Micelles
size and shape
        bile salt micelles, 81–84
        β-casein micelles, 86–88
        general considerations, 79–81
        phospholipid micelles, 84–86
Micellization
critical micelle concentration and, 64–67
enthalpy–entropy, 93–95
in mixed amphiphile systems, 95–98
multiple equilibrium model, 89–91
natural surfactants, 71–74
phospholipids, 74–75
protein associations similar to, 76–79
synthetic surfactants, 67–71
thermodynamics of, 89–95
Microtubule formation, 78–79
Milk proteins, 76–79
Mixed monolayers
alpha-helical conformation studies, 48–49
DPPC–phosphatidylglycerol, 42–43
        thermodynamic aspects of, 39–40
excess Gibbs energies in, 40–41
pH and formation, 49
phospholipid–cholesterol, 44–48
phospholipids, 41–44
spectroscopic examination, 48–49
surface pressure–isotherms, 41–44
surface pressures and surface potentials
        of DPPC-egg, 43
        of poly-L-glumatic acid, 49
Molecular dynamics
of biomembranes
of protein–surfactant complexes, 172–177
Molecular sieve chromatography, 154
Molecular weight, 80–81
Monolayers; *see also* Protein monolayers
diffusion studies, 233
equivalence with bilayers, 57–60, 58–60
fluorescent probe studies, 28–31
formation, 24–25
Gibbs, 24–25
Langmuir trough studies, 25–28
liquid interface studies, 25–29
mixed, 39–59; *see also* Mixed monolayers
phospholipid
        acyl chain length and temperature and,
                31–35
        dimyristoylphosphatidic acid (DMPA),
                37–39
        ion interactions and electrostatic
                effects, 36–39
        monolayer–surface bilayer transition,
                35–36

Monolayers (*Continued*)
    properties and chain melting, 108
    protein, 50–58
    states of, 29–31

NADH, 226
Natural surfactants, 8–10
    critical micelle concentration (cmc),
        71–74
Neonatal respiratory distress syndrome, 126
Neutron scattering, 154
N-linked glycoproteins, 184
NOE (nuclear Overhauser effect) studies, 195
Nonionic surfactants, 6
n regions, 20–21
Nuclear magnetic resonance (NMR)
    spectroscopy, 195, 222, 226

O-linked glycoproteins, 184
Onsager theory of conductivity, 65–66
Opsins, 196–197, 206, 207, 226, 238
Order profile parameter, 107
Overhauser effect studies, 195

PAGE analysis, 149–150, 154
Pearl-necklace model, 168–170
Penicillinase reconstitution study, 207
Pentadecanoic acid monolayer states, 30
Pfl coat protein studies, 195
pH
    and bile acid solubility, 73–74
    and chain melting, 112
    and liposomal targeting, 124–125
    and monolayer behavior, 36
Phagocytosis, and liposomal drug delivery,
    135–139
Pharmaceutical targeting, 124–139; *see also*
    Liposomal targeting
Phase behavior of phospholipids, 102–115,
    190–192
    cubic phases, 113–115
    hexagonal phases, 112
    lamellar phase, 103–110
    phase separation, 110–112
Phase separation, 110–112
Phosphatidylcholine. *See* Lecithin
Phosphatidylethanolamine, 186–187
Phosphatidylserine, 186–187
Phospholipid monolayer
    air–water interface studies, 32–35
    Ca^{2+} and, 36–39
    chain length/temperature variations, 31–35
    mixed, 39–49; *see also* Mixed monolayers
    pH and ionization, 36
    transition to surface bilayer, 35–36

Phospholipids; *see also* Lipids; Phospholipid
    monolayer
    acyl chains of, 10, 11
    critical micelle concentration (cmc),
        74–75
    and insulin transport, 213–214
    lyotropic mesomorphism, 102–115
    micellar size and shape, 84–86
    in monolayers, 31–39
    phase behavior
        cubic phases, 113–115
        hexagonal phases, 112
        lamellar phase, 103–110
        phase separation, 110–112
    phase transitions, 190–192
    polar head groups, 12
    specificity, 226–227
    stereospecific numbering of, 12–13
    surfactant interactions with, 144–146
    transporter function, 208–212
    X-group, 10, 11
Planar bilayer lipid membranes, 202
Plasma membrane. *See* Biomembrane
Polarity, and chain melting, 111
Polarity scales, 18–19
Polydispersity, and molecular weight, 80–81
Protein associations, micellar-like behavior
    of, 76–79
Protein monolayers, 50–58
    adsorption kinetics, 50–54
    surface layer structure, 54–58
Protein–surfactant complexes
    molecular dynamics, 172–176
    structure, 157–172
Proteins; *see also* Amino acids
    amphipathicity of, 13–17
    anchoring of, 20, 64, 200, 227–228
    band 3, 238
    biomembrane
        FTIR spectroscopy, 195–196
    biomembrane composition, 183–184, 187–
        188
    chaperon (heat shock), 222–224
    colicins, 224–225
    covalent attachment to fatty acids, 228–
        230
    denaturation, 150–153
    extrinsic *vs.* intrinsic, 187, 188, 227–228
    G, 239
    heat shock (chaperon), 222–224
    of human erythrocyte membrane, 193
    hydrophobic anchors, 20
    insertion of, 224–225
    integral membrane, 18
    lateral diffusion, 234–237

lipid–protein interactions, 225–227
membrane, 17–20
Pfl coat, 195
receptor, 189–190
reconstitution studies, 204–214
rotational diffusion, 237–238
SDS-PAGE analysis, 149–150
surfactant interactions
  binding and denaturation, 151–153
  denaturing and nondenaturing classes,
    150–151
  experimental techniques, 153–156
  principles, 149–150
  protein–surfactant complexes
    molecular dynamics, 172–176
    structure, 157–172
  thermodynamics, 156–167
surfactant-subunit complexes, 155
targeting domains, 20–21
translocation, 220–224, 221–224
Proteoliposomes, 127–134; *see also*
  Liposomal targeting

Quantitative equilibrium dialysis, 153–155

Receptor-mediated endocytosis, 239–242
Receptor processing, 239–242; *see also*
  Transmembrane signaling
Reconstitution
  cholesterol in, 214–217
  by complex formation, 202
  by insertion, 201–202
  membrane/glycoprotein systems, 204–217
  objectives, 199–200
  principles, 200–202
  by reverse of solubilization, 201
  signal transmission function of lipids,
    212–214
  surfactants for, 202–204
  transporter function of lipids, 208–212
RES. *See* Reticuloendothelial system
Reticuloendothelial system (RES), 124
  and liposomal drug delivery, 135–139
Reverse-phase evaporation method, 120
Rhodopsin, 196–197, 207, 226, 235, 237,
  238

Salting-out, 75
Saponins, 8–10
Sarcoplasmic reticulum, 225, 226
Saturation, and glucose transport, 210–212
Scatchard analysis, 157–161
SDS
  and apocytochrome c, 171–172

Gibbs energy of binding, 165
lysozyme binding in aqueous solution,
  152–153
micellization parameters, 94
Scatchard plots for, 160–161
thermodynamic parameters for lysozyme
  interaction, 166
SDS-PAGE analysis, 149–150, 154, 165–172
Semliki forest virus, 207
Signal peptides, 221–222
Signal recognition particles (SRPs), 221–222
Signal transmission, 189–190, 212–214
Singer equation, 55
Singer–Nicolson fluid-mosaic model, 185–186
Sodium dodecylsulphate. *See* SDS
Sodium taurocholate, micellization
  parameters, 94
Solubilization; *see also* Reconstitution
  reversal of, 199–217
  surfactant-induced, 147–149
Specificity, lipid, 226–227
Spectrin, 193–194
Spectroscopy
  of biomembranes, 195–197
  circular dichroism (CD), 195, 222
  electron spin resonance (ESR), 225, 227
  Fourier transform infrared (FTIR), 195–
    196, 222
  of mixed monolayers, 48–49
  nuclear magnetic resonance (NMR), 195,
    222, 226
  of signal sequences, 222
  ultraviolet difference, 154
  X-ray diffraction, 167–168, 195
Sphingolipids, 10, 11
Statistical effects on binding, 164–166
Stereospecific numbering, 12–13
Sucrose isomaltase reconstitution study, 207
Surface pressure
  β-casein, 55
  Ca^{2+} and
  mixed glycoprotein–DPPC, 57
Surface tension, and cmc determination,
  66–67
Surfactants
  anionic (natural), 6, 7, 150
  bile salt–related, 10
  biomembrane interactions
    with lipids, 144–146
    in membrane solubilization, 147–149
    principles, 143–144
  cationic, 6, 7–8
  CHAPS, 9–10
  chemical structures of synthetic, 6
  cholesterol-related, 8–10

Surfactants (*Continued*)
  critical micelle concentration (cmc)
    natural surfactants, 71–74
    principles and concept, 64–67
    synthetic surfactants, 67–71
  environmental effects, 98
  natural, 8–10, 71–74
  protein interactions
    binding and denaturation, 151–153
    denaturing and nondenaturing classes,
      150–151
    experimental techniques, 153–156
    principles, 149–150
    protein–surfactant complexes
      molecular dynamics, 172–176
      structure, 157–172
      thermodynamics, 156–167
  for reconstitution studies, 203–204
  single alkyl chain, 5–7
  synthetic (ionic), 5–8, 67–71, 150
Surfactant–subunit complexes, 155
Synthetic surfactants, 5–8
  critical micelle concentration (cmc),
    67–71

Targeting domains, 20–21, 221–222
Temperature
  and β-casein association, 86
  and chain melting, 32–35, 102–103, 105–
    112
  and cmc determination, 66–67
  and liposomal targeting, 124–125
  and monolayer/bilayer transition, 35–36
    cholesterol, 44–45
Thermodynamics
  of chain transitions, 105–107
  of micellization, 89–95
  of protein–surfactant interactions, 151–153
Thermotropic mesomorphism, 102–107
Titrimetry, 154

Transferrin, 239–240
Transition temperature, 102, 105
Transmembrane signaling, 189–190, 212–
  214, 239–242
Triton X-100, 150, 203–204
  micellization parameters, 94
  solubilization by, 146
  as solubilization marker, 146
Tubulin, 78–79
Tumor therapy, 124–125

UDP glucuronosyltransferase reconstitution
  study, 207
Ultracentrifugation, 154
Ultraviolet difference spectroscopy, 154
Unsaturuation, and glucose transport, 210–
  212
Urea, and surfactant cmc, 71

Vesicles; *see also* Liposomal targeting;
    Liposomes
  intermediate-sized unilamellar (IUVs), 118
  large unilamellar (LUVs), 118
  multilamellar (MLVs), 118
  small unilamellar (SUVs), 117–120
Virosomes, 126–127; *see also* Liposomal
    targeting
Viruses, Semliki forest, 207
Viscometry, 154
Voltage-gated ion channels, 189–190

Wyman's binding potential π concept, 162–
  164

X-group structures, 10, 11
X-ray diffraction spectroscopy, 155–156,
  167–168, 195

Zwitterionic surfactants, 6